Mathematical elements
of scientific computing

Mathematical elements of scientific computing

Ramon E. Moore

The University of Wisconsin

HOLT, RINEHART AND WINSTON, INC.
New York   Chicago   San Francisco   Atlanta
Dallas   Montreal   Toronto   London   Sydney

Copyright © 1975 by Holt, Rinehart and Winston, Inc.
All rights reserved

Library of Congress Cataloging in Publication Data

Moore, Ramon E.
Mathematical elements of scientific computing.

Bibliography
1. Electronic data processing — Numerical analysis.

I. Title.
QA297.M64   519.4   74-8185
ISBN 0-03-088125-0

Printed in the United States of America
5 6 7 8   038   1 2 3 4 5 6 7 8 9

To my parents, Edgar and Rose

Preface

This book is intended as a one-semester undergraduate text on computing methods for a student who has had a year or two of calculus and who has (or is just acquiring) a rudimentary knowledge of computer programming (for instance, in FORTRAN). This will include many sophomores and most juniors and seniors (and graduate students) in the sciences, engineering, or in mathematics. No previous course in linear algebra is assumed; the parts of the subject relevant to understanding the elements of scientific computing are developed in the text, including Appendix B on the more elementary aspects. At the same time, there is a fair amount of material that may be new even to many experts, especially the material on efficient use of Taylor series (Chapter 6) and on interval methods (Chapters 1, 3, 4, and 6).

The purpose of the book is to present an introduction to some of the principal sources of computational problems in science along with the simplest mathematical and computational techniques for solving them.

Considerable attention is paid to the method of *least squares*. In addition to the enormous range of its applicability as a curve-fitting technique in science, it constitutes a source of computational problems in linear algebra and in nonlinear systems. Its extensions include the elegant and very useful method of orthogonal projection onto a finite dimensional subspace of an infinite dimensional function space (Chapter 5).

The Exercises emphasize the development of experience in actually computing solutions to problems with the help of modern computers. It is assumed throughout that a machine is available and the whole slant of the text is on *how*, with the help of mathematical insight, to easily make use of such a powerful tool for the numerical solution of the various types of mathematical problems that occur in scientific work.

Local linearization (Chapters 3 and 5) and discretization (Chapter 6) are introduced as two of the most important mathematical elements of scientific computing. Newton's method is discussed (Chapter 3) both in its usual version for real and vector-valued functions and also in its interval version, which provides a simple method for obtaining bounds on zeros of functions.

Eigenvalues and eigenvectors are introduced in Chapter 7 in the natural setting of studying the asymptotic behavior of solutions of linear differential equations with constant coefficients (such as those which describe a tuned antenna circuit). In this way the student not only learns something about *how* to compute them but also *why* he should want to compute them in the first place.

A complete account is given in Appendix C of a real scientific computation that makes use of a number of the methods developed in the text. This computation constituted an independent check on the accuracy of a prediction of the general theory of relativity concerning the advance of the perihelion of Mercury beyond what is predicted by Newtonian gravitational theory.

A large body of mathematics and computing technique has arisen during the past few centuries from attempts to understand and predict the observed motions of heavenly bodies. More recently, this has included satellites and spaceships. It seemed natural, therefore, to include illustrations of computing methods for problems of this type.

I am indebted to Louis B. Rall for helpful suggestions and to John Slaugh for some of the exercises.

The University of Wisconsin   R. E. Moore
June 1974

Contents

Mathematical elements
of scientific computing

chapter 1

Approximate arithmetic

1.1 Introduction

The use of numbers for counting and for measuring is thousands of years old. While the integers are adequate for counting, it has long been known that not even ratios of integers are adequate for representing the result of measuring — at least not for the result of an ideal exact measurement. The length of the diagonal of a square with unit sides has long been known to be an *irrational* number. Pythagoras gave us the relation

$$a^2 + b^2 = c^2$$

for the lengths of the sides of a *right* triangle (see Figure 1.1). This relation

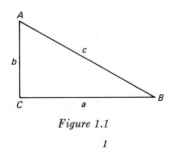

Figure 1.1

holds true only when the side AC is perpendicular to the side CB. If a and b are unit sides of a square and c is the length of the diagonal, then c must be a number such that

$$c^2 = 2$$

or as we now write

$$c = \sqrt{2}.$$

The following proof of the irrationality of this number was given by the ancient Greeks. It is a *proof by contradiction*. Suppose that there were integers m and n such that

$$c = \frac{m}{n}.$$

We can divide m and n by 2 a number of times until 2 no longer divides both m and n. One of them must then be odd (perhaps both). Squaring, we get

$$2 = \frac{m^2}{n^2}.$$

Multiplying both sides by n^2, we get

$$2n^2 = m^2.$$

It follows that m^2 must be divisible by 2, since the left-hand side of the equation is divisible and both sides are integers. But if m^2 is divisible by 2, then m itself must be divisible by 2, say

$$m = 2k.$$

We then have

$$2n^2 = 4k^2 \quad \text{or} \quad n^2 = 2k^2.$$

Repeating the argument just given, but with k and n in place of n and m, we can conclude that n must be divisible by 2. But we have now arrived at a contradiction to the previous statement that one of m and n must be odd. The conclusion from all this is that there cannot be integers m and n such that

$$\sqrt{2} = \frac{m}{n}.$$

This result implies, in particular, that $\sqrt{2}$ will have a nonrepeating decimal expansion.

EXERCISE

Show that any irrational number will have a nonrepeating decimal expansion.

1.2   Decimal Arithmetic and Digit-by-Digit Methods

Everyone is familiar with the convenience of the decimal form of numbers for the purpose of performing numerical calculations. The rules for getting arithmetic results digit-by-digit are few and simple. We can conveniently indicate the accuracy to which a result is known, furthermore, by reporting only as many decimal digits as are sure to be correct. While this convenience is available in the form of decimal numbers, however, it is not always so easy to know at the end of a long calculation how many digits are correct.

It is often no use to try to carry all digits that may occur during a lengthy computation, since the number of digits may grow so rapidly that the work will grind to a halt for lack of time. For example, the exact decimal representation of the tenth squaring of a three-digit number requires more than 3000 digits. If there were even the smallest fractional error in the third digit of the starting number, then we would not be justified in carrying very many digits in any case.

On the other hand, for a problem all of whose conditions are known exactly, it may be possible, in principle at least, to compute as many decimal digits as we like in the solution.

We can compute the decimal expansions of square roots in a *digit-by-digit* fashion.

The so-called *long-hand* method for this (similar to long division) amounts to the following: To obtain the successive digits $d_1, d_2, \ldots$ in

$$\sqrt{2} = 1.d_1 d_2 \cdots d_k \ldots,$$

suppose we already have

$$c_{k-1} = 1.d_1 d_2 \cdots d_{k-1};$$

we find the largest integer d_k less than or equal to 9 such that

$$r_k = 2 - c_k^2$$

is still positive, where

$$c_k = c_{k-1} + d_k \cdot 10^{-k}.$$

EXERCISE

Show that

$$r_k = r_{k-1} - .00\cdots d_k(2c_{k-1} + .00\cdots d_k),$$

as the long-hand method assumes.

We can describe the method just given for finding the first ten successive digits d_1, d_2, ... , d_{10} in the decimal expansion of $\sqrt{2}$ by the *flow chart* in Figure 1.2 as an aid to preparing a correct program.

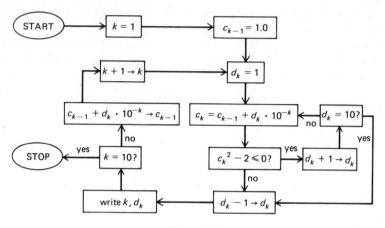

Figure 1.2

The notation $k + 1 \rightarrow k$ means "$k + 1$ replaces k"; $d_k + 1 \rightarrow d_k$ means "$d_k + 1$ replaces d_k," and so forth.

From $c_k = c_{k-1} + d_k \cdot 10^{-k}$, it follows that

$$c_k^2 = c_{k-1}^2 + 2c_{k-1}d_k \cdot 10^{-k} + d_k^2 \cdot 10^{-2k}.$$

If we define r_k, for $k = 0, 1, 2, \ldots$, by

$$r_k = 2 - c_k^2$$

then

$$r_k = r_{k-1} - d_k \cdot 10^{-k}(2c_{k-1} + d_k \cdot 10^{-k})$$

and we can see that this algorithm produces the same sequence of digits d_1, d_2, ... , as the so-called *long-hand* method previously described.

Let us write now a FORTRAN program for the alogorithm based on the flow chart description. We will attempt to use the program to find the first ten decimal digits in $\sqrt{2}$.

In the FORTRAN program we will use the following designations for variable names:

K	stands for k
CKM1	stands for c_{k-1}
CK	stands for c_k
IDK	stands for d_k (integer variable name)

The program may be written as follows:

C		FIND SQUARE R∅∅T OF 2
	1	K = 1
		CKM1 = 1.0
	2	IDK = 1
	3	CK = CKM1 + FL∅AT(IDK)*(10.**(−K))
	4	IF(CK**2 − 2.0) 41, 41, 42
	41	IDK = IDK + 1
		IF(IDK.EQ. 10) G∅T∅ 42
		G∅T∅ 3
	42	IDK = IDK − 1
	43	WRITE (6, 44) K, IDK
	44	F∅RMAT (10X, 2HK = ,I3, 5X, 4HIDK = ,I3)
	5	IF (K. EQ. 10) G∅ T∅ 60
		CKM1 = CKM1 + FL∅AT (IDK)*(10.**(−K))
		K = K + 1
		G∅ T∅ 2
	60	ST∅P
		END

We read the program as written (together with the proper control cards for the system monitor) into the Univac 1108 computer at the University of Wisconsin (hereinafter referred to as 1108UW); it was compiled by a FORTRANV compiler and executed. The resulting output was as follows:

K = 1	IDK = 4
K = 2	IDK = 1
K = 3	IDK = 4
K = 4	IDK = 2
K = 5	IDK = 1
K = 6	IDK = 3
K = 7	IDK = 6
K = 8	IDK = 2
K = 9	IDK = 9
K = 10	IDK = 9

Only the first six of these digits are correct. Why? It can be determined by carrying enough digits in the long-hand method that

$$\sqrt{2} = 1.414213562 \ldots$$

Why does the program "break down" at K = 7?

The answer lies in the fact that the evaluation of CK ** 2 − 2.0 (corresponding to $c_k^2 - 2$) is not carried out with enough precision.

Suppose $c_k = \sqrt{2} + e_k$. Then $c_k^2 = 2 + 2\sqrt{2}e_k + e_k^2$ and

$$c_k^2 - 2 = 2\sqrt{2}e_k + e_k^2.$$

Using single-precision arithmetic (see Section 1.4) the 1108UW cannot distinguish between two numbers that differ by less than about 1 in the 8th decimal digit. Counting the leading 1 in c_k and seven places after the decimal as 8 decimal digits, we should not expect the compiled program to be able to determine the correct sign (positive or negative) of $c_k^2 - 2$ when e_k is less than about 10^{-7}. This is what happened. In Section 1.4 we will discuss a means of guarding against computing errors of this kind.

1.3   Order Relations and Equivalence Relations

We can think of the so-called real numbers, integers, and rational numbers and irrational numbers as *points* laid out in an orderly fashion on a line with positive numbers to the right and negative numbers to the left (with apologies to those who are left-handed). We obtain in this way a one-dimensional continuum called the *real line*. This is nothing more or less than the x-axis in an x-y coordinate system. For any two real numbers a and b exactly one of the following relations is true:

1. $a < b$.
2. $a = b$.
3. $a > b$.

That is, a is less than b, equal to b, or greater than b. The first and third relations are examples of *order* relations, and the second is an example of an *equivalence* relation. An important property, in fact the *defining* property of order relations, is their *transitivity*, that is, *if $a < b$ and $b < c$, then $a < c$*. The relations $>$, $\leq$, $\geq$ also have this property and are order relations. Order relations (interesting ones, at least) are not *symmetric*, that is, if $a * b$ is true, then $b * a$ is not true, at least not for all a and b. The order relation $\leq$ is, of course, transitive, but it *is* symmetric for certain special pairs (a,b) namely when $a = b$.

Equality is an example of an *equivalence* relation, that is to say, a relation which is *transitive, symmetric*, and *reflexive*. Transitivity we have already discussed. A relation $a * b$ is *symmetric* if $a * b$ implies that $b * a$ also. For instance, $a = b$ implies that $b = a$. A relation $*$ is *reflexive* if $a * a$ for every a to which the relation applies.

Another example of an equivalence relation is afforded by *congruence* or *residue-class arithmetic* or *modular arithmetic*. A familiar example of this is in the determination of the time of day. We count the hours of the day *modulo* 12 (or *modulo* 24 in some places). A clock runs through a 12-hour cycle and then repeats its settings over again. If it is 6 o'clock now and we want to know what time it will be 100 hours from now, we add to get 106; then we write 106 as some multiple of 12 plus a residue or remainder *less than 12*. Thus we obtain

$$106 = (8)(12) + 10 \quad \text{or} \quad 106 \equiv 10 \text{ (modulo 12)}.$$

We can use the special symbol $\equiv$ to represent the equivalence relation defined by arithmetic modulo some fixed number N. Thus, after reaching an agreement on what N we are talking about, we can write $a \equiv b$ whenever $a - b$ is exactly divisible by N. To keep track of N, we could use the safer notation $a \equiv b(N)$.

Thus we have, for example, $13 \equiv 1(12)$, $16 \equiv 4(12)$, $1 \equiv 1(12)$, $12 \equiv 0(12)$, and so forth. The relation $\equiv$ is transitive, since $a - b$ being divisible by N and $b - c$ being divisible by N implies that $a - c = (a - b) + (b - c)$ is also divisible by N. Being divisible by N, by the way, means that the quotient upon division by N is an integer. The relation is reflexive since $a - a$ is always zero and gives an integer quotient, namely zero, upon division by N. The relation is symmetric since $a - b$ is divisible by N whenever $b - a$ is divisible by N.

If N is a positive integer, then $a \equiv b(N)$ makes every integer equivalent (with respect to this congruence) to one of the integers: $0, 1, 2, \ldots, N - 1$.

We can also use this sort of relation when N is not an integer and, for instance, we can define $a \equiv b(N)$, whenever $(a - b)/N$ is an integer. This will make every real number equivalent to some number in the *interval* from 0 to (but not including) N. Such a thing is useful, for instance, when dealing with periodic functions. For instance, the sine and cosine functions have period 2π. This is to say that

$$\sin (x + k2\pi) = \sin (x) \quad \text{and} \quad \cos (x + k2\pi) = \cos (x)$$

for every real number x and for any integer k. Thus to compute a value for $\sin (y)$ for some large argument y, we can first find x between 0 and 2π such that $y \equiv x(2\pi)$ (read "y equals x mod 2π") and then compute $\sin (y) = \sin (x)$. In order to do this, we divide y by 2π and then throw away the integer part of the quotient and save the remainder! Actually, if $y/2\pi = k + r$ with $0 < r < 1$, then $x = y - k \cdot 2\pi = 2\pi r$. For example,

$$\sin (15.1) = \sin (15.1 - 4\pi) \approx \sin (2.53).$$

EXERCISES

1. Which of the following relations are *order relations* (transitive) and which are *equivalence relations* (reflexive, symmetric, *and* transitive)?
    (a) votes the same way as

(b) $<$
(c)   is a brother of
(d) $\leq$
(e) $=$
(f)   is geometrically similar to
(g) $x \equiv y$ (modulo n)
(h)   is greater than
(i)   is hotter than
(j)   is worse than
(k)   is a divisor of
(l)   is twice as large as
(m) is faster than
(n)   is approximately equal to

2. Represent the letters A to Z by the integers 1 to 26. Write an algorithm (for example, a FORTRAN program) for rearranging any given list of words into *alphabetical (lexicographic) order*.

3. Give two different solutions for a pair of integers x and y such that $3x + 2y \equiv 1 \pmod 7$.

4. Express in simplest form:
   (a) 10 (mod 3)
   (b) 143 (mod 12)
   (c) 39 (mod 3)
   (d) 1 (mod 2)
   (e) $\dfrac{7\pi}{2}$ (mod 2π)
   (f) $\cos 6\pi$

5. Write a computer program to convert an angle (ANGLE, expressed in degrees in decimal form) to three separate integers (IDEGS, MINS, and ISECS) representing corresponding degrees, minutes, and seconds, correct to the nearest second. For instance, if ANGLE = 43.626, your program should produce IDEGS = 43, MINS = 37, and ISECS = 34.

1.4   Interval Arithmetic and Error Analysis

We most often begin a scientific computation with inexact initial data. We wish to determine approximate values of certain quantities from formulas or equations that will usually be only approximate descriptions of the *natural* relations between the quantities involved. Often we must use *approximate arithmetic*, carrying only a certain number of decimal (or binary) digits because of the inefficiency of attempting to carry all digits (and, anyway, division produces infinite decimal expansions; for instance,

$\frac{1}{3} = .333333\ldots$). Many special techniques are available for analyzing and estimating the effects of the various approximations made in the course of a computation. Some of these will be discussed in the following sections. There is a simple, general technique, called *interval arithmetic*, for keeping track of the effects of using approximate arithmetic and of beginning with approximate initial data that can be discussed at this point. We can begin with an example to illustrate the idea.

The formula (see Appendix on "closed-form solutions")

$$V = \sqrt{\frac{2gM}{E(M + E)}} - V_0$$

gives an approximation to the minimum excess velocity over the earth's velocity in its orbit around the sun which a spaceship must have in power-less flight after it leaves the strong influence of the earth's gravitational field in order to reach the vicinity of Mars some six months later. The terms in the formula may be assumed here to be known within the following degrees of approximation:

$$4.1 \cdot 10^{17} < g < 4.2 \cdot 10^{17}$$
$$66 \cdot 10^3 < V_0 < 67 \cdot 10^3$$
$$141 \cdot 10^6 < M < 143 \cdot 10^6$$
$$92 \cdot 10^6 < E < 94 \cdot 10^6.$$

The physical meaning of the terms and their units are as follows: g is a constant describing the sun's gravitational attraction and is given in miles^3/hour^2; V_0 represents in miles per hour the earth's velocity in its orbit around the sun; M is the radius of the orbit of Mars around the sun in miles; and E is the radius of the earth's orbit in miles.

Another way of denoting the above set of limiting values would be

$$g = (4.15 \pm .05) \cdot 10^{17}$$
$$V_0 = (66.5 \pm .5) \cdot 10^3$$
$$M = (142 \pm 1) \cdot 10^6$$
$$E = (93 \pm 1) \cdot 10^6.$$

If we denote the *interval* of real numbers between a and b by the symbol $[a,b]$, then we can give the above data in still another way, namely:

g is in the interval $[4.1,4.2] \cdot 10^{17}$.

V_0 is in the interval $[66,67] \cdot 10^3$.

M is in the interval $[141,143] \cdot 10^6$.

E is in the interval $[92,94] \cdot 10^6$.

The expression $[4.1,4.2] \cdot 10^{17}$ means the same thing as $[4.1 \cdot 10^{17}, 4.2 \cdot 10^{17}]$.

We will now evaluate the formula for V using *interval arithmetic*, beginning with the interval form of the inexact initial data. It will be seen that *during the computation* we can keep track of the inexactness in initial data as well as the effects of dropping decimal digits (or rounding off, as we prefer).

We will use the notation "$x \in I$" for "x is in the interval I." It can be seen that

$$2g \in [8.2, 8.4] \cdot 10^{17}$$

and

$$2gM \in [(8.2)(141), (8.4)(143)] \cdot 10^{23}.$$

Now $(8.2)(141) = 1156.2$ and $(8.4)(143) = 1201.2$; so, carrying three digits, we have

$$2gM \in [11.5, 12.1] \cdot 10^{25}.$$

It is also true that

$$2gM \in [11.562, 12.012] \cdot 10^{25},$$

but there seems little point in carrying along all those extra digits through the rest of the computation, since the *right* and *left end-points* of the interval [11.562,12.012] already differ by about 5 percent.

Next, we can compute that

$$M + E \in [233, 237] \cdot 10^{6}$$

and

$$E(M + E) \in [(92)(233), (94)(237)] \cdot 10^{12}.$$

Since $(92)(233) = 21,436$ and $(94)(237) = 22,278$, we can deduce that

$$E(M + E) \in [21.4, 22.3] \cdot 10^{15}.$$

Using the result, computed earlier, that

$$2gM \in [11.5, 12.1] \cdot 10^{25},$$

we can compute that

$$\frac{2gM}{E(M + E)} \in \left[\frac{11.5}{22.3}, \frac{12.1}{21.4} \right] \cdot 10^{10}.$$

Carrying out the indicated divisions, we obtain

$$\frac{11.5}{22.3} = .5156 \ldots \quad \text{and} \quad \frac{12.1}{21.4} = .5654 \ldots$$

so that we can conclude

$$\frac{2gM}{E(M + E)} \in [.515, .566] \cdot 10^{10}.$$

Now, if $x \in [a,b]$ with a and b positive numbers such that $a < b$, then

$$\sqrt{x} \in [\sqrt{a}, \sqrt{b}];$$

therefore we compute

$$\sqrt{.515} = .717 \ldots \quad \text{and} \quad \sqrt{.566} = .752 \ldots$$

and conclude that

$$\sqrt{\frac{2gM}{E(M+E)}} \in [.717, .753] \cdot 10^5.$$

Finally, we compute that

$$V = \sqrt{\frac{2gM}{E(M+E)}} - V_0 \in [.717, .753] \cdot 10^5 - [66,67] \cdot 10^3$$

and so

$$V \in [71.7 - 67, 75.3 - 66] \cdot 10^3.$$

In other words, under the assumptions implied by the form of the data and the formula for V, we can conclude that $V \in [4.7, 9.3] \cdot 10^3$ or the minimal excess velocity V must under these assumptions be something between 4700 mph and 9300 mph. Sharper estimates of V would require more accurate knowledge of g, V_0, E, and M.

For instance, if we were to *assume* that $g = 4.15 \cdot 10^{17}$, $V_0 = 66.5 \cdot 10^3$, $M = 142 \cdot 10^6$, $E = 93 \cdot 10^6$, then we could compute to three places that

$$V = 10^5 \sqrt{\frac{11,786}{21,855}} - 66.5 \cdot 10^3 = 6.93 \cdot 10^3$$

or about 6930 mph.

The general rules for determining end-points during interval arithmetic operations are as follows:

$$[a,b] + [c,d] = [a + c, b + d]$$
$$[a,b] - [c,d] = [a - d, b - c]$$
$$[a,b] \cdot [c,d] = [\min (ac,ad,bc,bd),$$
$$\max (ac,ad,bc,bd)]$$
$$[a,b] / [c,d] = [a,b] \cdot [1/d, 1/c]$$
$$(\text{only defined if } 0 \text{ is not in } [c,d])$$
$$-[a,b] = [-b, -a]$$
$$c[a,b] = [a,b]c = [ac,bc] \quad (\text{if } c > 0).$$

EXERCISE

Find upper and lower bounds to the range of values of the polynomial

$$p(x) = x^7 + x^3 - 6x^2 + .11x - .006$$

when $0 \leq x \leq .2$ (substitute the interval $[0,.2]$ for x and evaluate the polynomial using interval arithmetic).

Further rules can be given for more complicated interval computations and one was given earlier for square roots.

The multiplication of intervals was given in terms of the *mini*mum and *maxi*mum of four products of end-points. Actually, by testing for the signs (positive or negative) of the end-points a, b, c, d, the formula for the end-points of the interval product can be broken into nine special cases. In eight of these cases only two products need be computed. (For instance, if $0 < a < b$ and $0 < c < d$, then $[a,b] \cdot [c,d] = [ac,bd]$.) In one case, when $a < 0 < b$ and $c < 0 < d$, all four products, (ac,ad,bc,bd), are needed since in this case $[a,b] \cdot [c,d] = [\min \ (ad,bc), \max \ (ac,bd)]$.

Discussion of Rounded Interval Arithmetic Implementation in FORTRAN

It is possible to secure complete protection against the sort of uncertainty which occurred in the computer results described in Section 1.2 by programming computations in *rounded interval arithmetic*.

The "uncertainty" referred to here concerns the question of how many decimal digits in the results of a machine computation will be correct when, at each stage, the computations are only carried out to a limited precision, (that is, to a fixed number of decimal or binary digits).

We can modify the rules given previously in this section for determining end-points during interval arithmetic operations so that by appropriate *rounding*, the exact result will be *contained in* the machine-computed, *rounded interval* result. Suppose that the machine *single-precision, floating-point* arithmetic operations act as follows (appropriate changes can be made for other specific versions of machine arithmetic).

Suppose that *single-precision** (floating-point) machine numbers have the form

$$\pm .d_1 d_2 \ldots d_n \cdot 10^e \quad (d_1 \neq 0 \text{ or else } d_1 = d_2 \ldots = d_n = 0)$$

where $d_1, d_2, \ldots, d_n$ are digits between 0 and 9, there is a sign: $+$ or $-$, and e is an integer (positive, zero, or negative).

The *exact* addition, subtraction, multiplication, or division of two such numbers would yield a result that could be represented as

$$\pm \ (.r_1 r_2 r_3 r_4 \ldots) \cdot 10^f$$

for some decimal digits $r_1, r_2, \ldots$ and some sign $(+$ or $-)$ and some integer f.

---

*Double-precision machine numbers will then have about $2n$ digits; N-tuple precision numbers will have about Nn digits. For binary arithmetic, replace 10 by 2 and "between 0 and 9" by "0 or 1" in the following discussion.

Suppose that the corresponding result of a machine arithmetic operation yields

$$\pm (.r_1r_2\cdots r_n)\cdot 10^J.$$

Thus we suppose that the machine result is simply the exact result *chopped off* after the nth decimal place (properly scaled). (Again we suppose that $r_1 \neq 0$ or else $r_1 = r_2 = \cdots = r_n = 0$.)

For rounded interval arithmetic on such a machine, we can use the following rules:

1. If $+ (.r_1r_2\cdots r_n)\cdot 10^J$ represents the machine arithmetic computation of the *right-hand* end-point of an interval, then take

$$+ (.r_1r_2\cdots r_n)\cdot 10^J(1 + 10^{-n+1})$$

   as the *rounded interval arithmetic* right-hand end-point.

2. If $+(.r_1r_2\cdots r_n)\cdot 10^J$ represents the machine arithmetic computation of the *left-hand* end-point of an interval, then take $+(.r_1r_2\cdots r_n)\cdot 10^J$, itself as the *rounded interval arithmetic* left-hand end-point.

3. If $- (.r_1r_2\cdots r_n)\cdot 10^J$ represents the machine arithmetic computation of the *right-hand* end-point of an interval, then take $- (.r_1r_2\cdots r_n)\cdot 10^J$, itself, as the *rounded interval arithmetic* right-hand end-point.

4. If $- (.r_1r_2\cdots r_n)\cdot 10^J$ represents the machine arithmetic computation of the *left-hand* end-point of an interval, then take $- (.r_1r_2\cdots r_n)\cdot 10^J(1 + 10^{-n+1})$ as the *rounded interval arithmetic* left-hand end-point.

In the special case (in rule 1 or rule 4) that $r_1 = r_2 = \cdots r_n = 9$, take $\pm(1.000)\cdot 10^{J+1}$ as the rounded interval arithmetic end-points.

We will illustrate the application of these rules to the algorithm discussed in Section 1.2. What we propose to do is to replace the machine arithmetic evaluation of the expressions

$$c_k = c_{k-1} + d_k\cdot 10^{-k}$$

and $c_k^2 - 2$, which occurred in the algorithm discussed in Section 1.2, by rounded interval arithmetic evaluations of these expressions, using $c_k = [a_k,b_k]$. Furthermore, we will replace the test

$$\boxed{c_k^2 - 2 \leq 0?}$$

by its rounded interval arithmetic version.

Suppose $[A,B]$ is the *rounded* interval arithmetic result corresponding to $c_k^2 = [a_k,b_k]^2$; then the *exact* value of c_k^2 is contained in the interval $[A,B]$ and the following *three* things can happen:

1. $[A,B] - 2 \leq 0$; that is, $B - 2 \leq 0$, which means that

$$c_k{}^2 - 2 \leq 0 \text{ is } \textit{definitely true;}$$

2. $0 < [A,B] - 2$, that is, $0 < A - 2$, which means that

$$0 < c_k{}^2 - 2 \text{ is } \textit{definitely true;} \quad \text{or}$$

3. $A - 2 \leq 0 < B - 2$, in which case it is *uncertain* whether

$$c_k{}^2 - 2 \leq 0 \quad \text{or} \quad 0 < c_k{}^2 - 2.$$

As long as (1) or (2) occurs we can continue to determine the successive digits d_k with certainty. As soon as (3) occurs, we must stop the computation because we have exceeded the *limits of machine precision.*

We will denote *rounded interval addition by* $\oplus$, *rounded interval subtraction* by $\ominus$, and *rounded interval multiplication by* $\odot$. We can now revise the flow chart given in Figure 1.2 to the rounded interval arithmetic version, as shown in Figure 1.3. Notice that the main difference between this and the flow chart in Figure 1.2 is that we now *stop* the algorithm *before an uncertainty occurs* in any result.

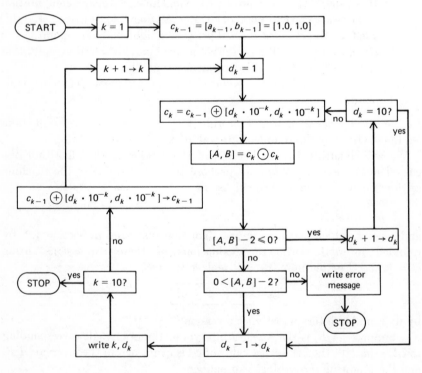

Figure 1.3

Let us now write a FORTRAN program for the rounded interval arithmetic version of the algorithm as described by the flow chart in Figure 1.3. We will use the following designations for variable names:

K   stands for k

AKM1   stands for a_{k-1}

BKM1   stands for b_{k-1} where $c_{k-1} = [a_{k-1}, b_{k-1}]$

AK   stands for a_k

BK   stands for b_k where $c_k = [a_k, b_k]$

IDK   stands for d_k

A   stands for the left end-point of $c_k \odot c_k$

B   stands for the right end-point of $c_k \odot c_k$

The program may be written as follows (we may use $n = 8$ for the 1108UW):

C			
			FIND SQUARE RØØT OF 2
	1		K = 1
			AKM1 = 1.0
			BKM1 = 1.0
	2		IDK = 1
	3		AK = AKM1 + FLØAT(IDK)*(10.**(−K))
			BK = (BKM1 + FLØAT(IDK)*(10.**(−K)))*(1. + 10.**(−7))
			A = AK**2
			B = (BK**2)*(1. + 10.**(−7))
	4		IF (B − 2.) 41, 41, 40
	41		IDK = IDK + 1
			IF (IDK.EQ.10) GØ TØ 42
			GØ TØ 3
	40		IF (A − 2.) 61, 61, 42
	42		IDK = IDK − 1
	43		WRITE (6, 44) K, IDK
	44		FØRMAT(10X, 2HK = ,I3, 5X, 4HIDK = ,I3)
	5		IF (K.EQ. 10) GØ TØ 60
			AKM1 = AKM1 + FLØAT(IDK)*(10.**(−K))
			BKM1 = (BKM1 + FLØAT(IDK)*(10.**(−K)))*(1.
		1	+ 10.**(−7))
			K = K + 1
			GØ TØ 2
	60		STØP
	61		WRITE (6, 62)
	62		FØRMAT(10X, 26H LIMIT ØF MACHINE PRECISIØN)
			STØP
			END

We read the program as written (with the proper control cards) into the 1108UW. It was compiled and executed. The resulting output was as follows:

K = 1   IDK = 4
K = 2   IDK = 1
K = 3   IDK = 4
K = 4   IDK = 2
K = 5   IDK = 1
LIMIT OF MACHINE PRECISION

By taking into account the *binary* representation of machine arithmetic on the 1108UW we can justify using the factor $(1 + 2^{-26})$ instead of $(1 + 10^{-7})$ in statements $3 + 1$, $3 + 3$, and $5 + 2$ in the above program. With these changes we obtain finally the following *perfect* output (see Section 1.2):

K = 1   IDK = 4
K = 2   IDK = 1
K = 3   IDK = 4
K = 4   IDK = 2
K = 5   IDK = 1
K = 6   IDK = 3
LIMIT OF MACHINE PRECISION

EXERCISES

1. If $a \in [1,2]$ and $b \in [3,4]$, find intervals containing $a + b$, $a - b$, ab, and a/b.

2. Find intervals containing $a + b$, $a - b$, and ab for each of the following examples:
    (a) $a \in [-1,2]$, $b \in [2,3]$
    (b) $a \in [-2,1]$, $b \in [-1,1]$
    (c) $a \in [-2,-1]$, $b \in [1,5]$
    (d) $a \in [-2.1,-2]$, $b \in [-1,0]$

3. Suppose we know only that

$$x \in [7.853, 7.854].$$

Find $[a,b]$ such that $\sin x \in [a,b]$. *Hint:* Restrict π to a known interval, and to evaluate the sine of an angle use $\sin y = \cos [(\pi/2) - y]$ and

$$\cos y = 1 - \frac{y^2}{2} + \frac{y^4}{24} + \cdots.$$

4. Write a computer program for carrying out rounded interval multiplication, making appropriate sign tests so that only two products need be computed in most cases.

1.5 Complex Arithmetic

Our discussion of the types of numbers used in computing must include mention of *complex* numbers. Just as the *real* numbers can be visualized as points laid out in a line ordered by the relation $<$, so the complex numbers can be visualized as points in the plane. Given the Cartesian coordinates (x,y) of a point in the plane, the *ordered pair* of real numbers (x,y) is a representation of a *complex number*. The components x and y are called, respectively, the *real part* and the *imaginary part* of the complex number (x,y). This terminology is unfortunate, since both x and y are *real* numbers! It is too bad that these components of a complex number were not dubbed something like: the *first* part and the *second* part, or the left part and the right part; or the x part and the y part; anything but real and "imaginary"!

Anyway, arithmetic operations are defined for complex numbers as follows:

$$(x_1,y_1) + (x_2,y_2) = (x_1 + x_2, y_1 + y_2)$$
$$(x_1,y_1) - (x_2,y_2) = (x_1 - x_2, y_1 - y_2)$$
$$(x_1,y_1) \cdot (x_2,y_2) = (x_1 x_2 - y_1 y_2, x_1 y_2 + x_2 y_1)$$
$$\frac{(x_1,y_1)}{(x_2,y_2)} = \left(\frac{x_1 x_2 + y_1 y_2}{x_2^2 + y_2^2}, \frac{x_2 y_1 - x_1 y_2}{x_2^2 + y_2^2} \right).$$

The number $(x,-y)$ is called the *complex conjugate* of the number (x,y). We can also use the notation $\overline{(x,y)} = (x,-y)$.

These definitions for complex arithmetic operations give us an arithmetic for points in the plane. When viewing points in the plane as complex numbers in this way, we refer to the plane as the *complex plane*.

If we denote the complex number (x,y) by a single symbol, say z, then we can write the complex conjugate of z as $\bar{z}$. Now for any complex number $z = (x,y)$ we will have $z\bar{z} = (x,y)(x,-y) = (x^2 + y^2, 0)$. We can identify points on the x-axis, which are complex numbers of the form $(x,0)$, with ordinary real numbers. Thus $(x,0) \equiv x$ is an equivalence relation. Having done this, we can now write $z\bar{z} = x^2 + y^2$. Thus $z\bar{z}$ is always a nonnegative real number. In fact, $\sqrt{z\bar{z}} = \sqrt{x^2 + y^2}$ is the distance of the point $z = (x,y)$ from the origin of the Cartesian coordinate system for the plane. We call $|z| = \sqrt{z\bar{z}}$ the *magnitude* of z.

We can now describe the unit circle, as shown in Figure 1.4 (the circle with center at the origin and with unit radius), by the simple equation

$$|z| = 1$$

which is the same as

$$x^2 + y^2 = 1.$$

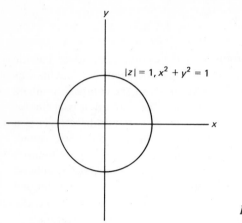

$$|z| = 1, x^2 + y^2 = 1$$

Figure 1.4

Referring to the definition of multiplication for two complex numbers, we can verify that the complex numbers $(0,1)$ and $(0,-1)$ both satisfy the equation

$$z^2 + 1 = 0.$$

For instance, if $z = (0,1)$, then

$$z^2 = z \cdot z = (0,1) \cdot (0,1) = (-1,0) = -1.$$

Another notation for a complex number z in terms of its components is the following. Just as we denoted, in Section 1.1, the positive irrational number c which satisfies the equation

$$c^2 = 2$$

by the symbol $\sqrt{2}$, so we could denote the complex number $z = (0,1)$ which satisfies the equation

$$z^2 = -1$$

by the symbol $\sqrt{-1}$. Other symbols for the complex number $(0,1)$ are the letter i (commonly used in mathematical writings) or the letter j (commonly used in engineering writings).

Using the various notations that have been mentioned, we can write any complex number $z = (x,y)$ in the alternative forms:

$$z = x + (0,1)y;$$
$$z = x + \sqrt{-1}y;$$
$$z = x + iy;$$
$$z = x + jy.$$

EXERCISES

1. Using *complex arithmetic*, find $z + w$, $z - w$, zw, and z/w for the following *pairs* of complex numbers z,w:
    (a) $z = (1,1)$, $w = (1,-1)$.
    (b) $z = (2,0)$, $w = (2,0)$.
    (c) $z = (0,2)$, $w = (0,2)$.
    (d) $z = (1,1)$, $w = (2,2)$.
2. Compute and plot the points (complex numbers) z^2, z^3, z^4 when

$$z = \left(\frac{1}{\sqrt{2}}, \frac{1}{\sqrt{2}}\right).$$

3. If $z = (x,y)$ is a point in the complex plane (complex number), let

$$x = r \cos \theta$$
$$y = r \sin \theta.$$

Express r and θ in terms of x and y. Give a geometric interpretation of the *polar coordinates* r, θ for the point z.
4. Write a computer program to find the polar coordinates of a complex number $z = (x,y)$.

1.6   Rounding Errors

We use the terms *rounding errors* or *round-off errors* to describe the differences between exact (or *infinite precision*) arithmetic and approximate (or *finite precision*) arithmetic.

Computations carried out in rounded interval arithmetic will produce intervals containing both the exact (infinite precision) results and also the ordinary machine arithmetic results. Therefore interval arithmetic can provide strict bounds on the accumulated rounding error (as well as propagated initial error) in any machine computation.

It is not difficult to write computer subroutines for carrying out rounded interval arithmetic. This can be done in a variety of ways; see References at end of book (Apostolatos et al., 1968; Boche, 1963; Collins, 1960; Gibb, 1961; Good, 1968; Ladner and Yohe, 1970; Moore, 1962 and 1966; Nickel, 1968 and 1969; Reiter, 1968; Richman, 1969).

Several questions naturally arise:

   1. Are the bounds realistic?

    2. What is the cost in extra computing time of obtaining interval bounds?

    3. Is there a way of computing with statistical estimates or confidence intervals instead of strict bounding intervals?

For the first question, we will now consider some examples of the actual performance of interval arithmetic routines on various problems and on various computers. (See also Section 4.6.1.)

EXAMPLES

1. The recurrence formula

$$I_n = 1 - nI_{n-1}, \qquad n = 1, 2, 3, \dots$$

starting with $I_0 = 1 - 1/e = .63212\ 05588 \dots$ yields values of

$$I_n = \frac{1}{e} \int_0^1 x^n e^x dx$$

(Babuska, et al., 1966; Moore, 1968). Using rounded interval arithmetic on the CDC1604 (Reiter, 1968), starting with $I_0 = [.63212\ 05588, .63212\ 05589]$, the intervals I_n were computed for $n = 1, 2, \dots, 14$ (Moore, 1968, p. 60). In particular, for $n = 10$, the interval

$$I_{10} = [.0838366, .0842350]$$

was obtained. Note here that the width of I_0 was 10^{-10} and we can show that this alone accounts for most of the error propagation. Denote the width of I_n (in the interval arithmetic evaluation of the recurrence formula) by $w(I_n)$; then we have

$$w(I_n) = nw(I_{n-1})$$

even without any further rounding error so that the error 10^{-10} in the initial data propagates to $10!10^{-10} \approx 3.6 \cdot 10^{-4}$ for $n = 10$. For $n = 10$, the interval I_{10} actually computed had width $3.98 \dots \cdot 10^{-4}$. These bounds *are* realistic estimates of actual accumulated computational error.

2. (See Babuska et al., 1966; Moore, 1968.) Again using rounded interval arithmetic on the CDC1604 (Reiter, 1968) we programmed the sequence $y_n = z_n/n$, $z_{n+1} = ny_n$, $n = 1, 2, \dots, 10000$ with $z_1 = 1$. The computer obtained the result

$$y_{10000} \cdot 10^4 \in [.99999\ 9621534, 1.00000024744].$$

That this, too, is realistic can be argued as follows. For the *same* calculation, but using *ordinary machine* (floating-point) arithmetic, a variety of different results were obtained at several computing centers in Europe, as reported by Babuska et al., 1966, p. 7.

Computer	Computing Center Located in	$y_{10000} \cdot 10^4$	Comments
ZUSE 23	Prague	0.99999 3645	
LGP 30	Brno	0.9992631	Machine language
LGP 30	Prague	0.9995524	Automatic coding
LGP 30	Prague	1.001660	Special subroutine
D2	Dresden	1.00000000002	
X1	Braunschweig	1.00000000000	12 dec. places, aut. coding
X1	Braunschweig	0.99996060	8 dec. places, aut. coding
SIEMENS 2002	Mainz	0.999999761	
ER 56	Stuttgart	0.99999972512	

3.   (See Nickel, 1968.) The series

$$\sum_{i=1}^{10000} \left(\frac{(i-1)(i-2)(i^2-2)}{(i+1)(i+2)(i^2+2)} - 1 \right)^2$$

was computed on the IBM 360/40 at the IBM research laboratory in Zurich using rounded interval arithmetic and programmed in PL/1. The resulting interval was

$$[8.578457069097124, \ 8.578457069099536].$$

In other words, taking the midpoint of this interval as an approximate value for the series, we have an error bound for that value of about 10^{-12}. The interval calculation has bounded the round-off-error accumulation at about 1 in the fourth to last place.

Let us consider now the *second* question raised: "What is the cost in extra computing time of obtaining interval bounds?"

Collins (1960) finds that a *ratio* (for total computing time of an interval computation divided by total computing time of the same computation in ordinary machine arithmetic) "might typically be of the order of 4 or 5"; this is based on a ratio of about 12 to 1 for his individual interval arithmetic operations. Boche (1963) conjectures a minimum ratio of around 2 to 1 for individual arithmetic operations, which would drop Collins' estimate for the overall ratio to an increase of 70 to 80 percent. At any rate, something like a doubling of computing time would seem a minimum estimate for an ideal computer software realization. If interval arithmetic operations were made into hardware features of a computer, then this figure would decrease. Boche's interval arithmetic subroutines for the IBM 7094 (Boche, 1963) had the following execution times (in microseconds) compared to single-precision machine arithmetic:

	Add and subtract	Multiply	Divide
Interval	84	60	60
Single precision	22	8	16

As far as the *third* question goes, *statistical estimation* of round-off error accumulation has been studied for a variety of computational problems (Chai, 1967; Dempster, 1969; Henrici, 1962). It is possible to obtain smaller estimates of error in this way than by using interval arithmetic. On the other hand, it may be difficult or impossible to interpret precisely the *meaning* of probability statements in error estimates based on assumptions of independence that are generally not true in practice during the course of a particular computation (Chai, 1967). As a result, a statistical approach may overestimate or underestimate the actual error by amounts that are difficult to predict.

An approach to the analysis of rounding error that is especially useful for computational problems in linear algebra is the so-called *inverse-error analysis* or *backward-error analysis* of Givens, 1954 and Wilkinson, 1963. An example of this would be the following.

Suppose we are given a matrix A and a vector b and wish to find x such that (see Chapter 4)

$$Ax = b$$

If we employ some particular algorithm and finite precision arithmetic to find an approximate solution $\bar{x}$, then we may ask ourselves: What matrix δA would we have to add to A so that $\bar{x}$ is an *exact* solution to

$$(A + \delta A)\bar{x} = b?$$

Wilkinson (1963) has derived bounds on the norm of δA for $\bar{x}$ determined by Gaussian elimination with pivoting. Forsythe and Moler (1967) represent these bounds in the form

$$\|\delta A\|_\infty \leq 1.01 \ (n^3 + 3n^2)\rho \ \|A\|_\infty u,$$

where

$$\|\delta A\|_\infty = \max_{i=1,\,2,\,\ldots,\,n,} \sum_{i=1}^{n} |\delta A_{ij}|$$

is the maximum row-sum norm of δA, n is the order of A, u is a unit round-off error, and ρ is a bound on quantities which occur during the elimination process

$$\rho = \max_{i,j,k} |a_{i,j}^{(k)}| \ \|A\|_\infty$$

where $A^{(k)} = (a_{i,j}^{(k)})$ is zero below the diagonal in the first $k - 1$ columns. But here again, things usually go better than the bounds indicate. Evidently $\|\delta A\|_\infty$ is, in practice, "rarely larger than $nu \ \|A\|_\infty$" (Forsythe and Wasow, 1960).

Both interval methods (Moore, 1966) and statistical methods (Henrici, 1962) have been applied to the analysis of rounding error in computational methods for differential equations. A variety of additional approaches are under development including variable precision-interval arithmetic (Richman, 1972) and ellipsoidal bounding procedures (forthcoming work by W. Kahan).

For more extensive lists of references on the whole subject of numerical error analysis, see the two volume work, edited by L. B. Rall, 1965.

chapter

Recursion and iteration

2.1 Introduction

Several hundred years ago the astronomer Kepler calculated an elliptical orbit for the planet Mars from observations of the angular position of Mars at various times. He used what was essentially an *iterative* method and spent five years on these calculations. He left careful records of his work and the calculations were recently repeated in a few minutes on a high-speed computing machine (Gingerich, 1964).

In this chapter we introduce some basic notions concerning recursion formulas and iterative methods. We will expand on these in subsequent chapters.

How can we describe what it is that we do in multiplying two numbers in decimal form? We can give an algorithm for the way we perform such an operation, with *carries*, and so on, as follows. Let

$$N = d_n \cdot 10^n + d_{n-1} \cdot 10^{n-1} + \cdots + d_1 \cdot 10 + d_0 \qquad (0 \leq d_i \leq 9, d_n \geq 1)$$

and

$$M = D_m \cdot 10^m + D_{m-1} \cdot 10^{m-1} + \cdots + D_1 \cdot 10 + D_0$$
$$(0 \leq D_i \leq 9, D_m \geq 1).$$

To find the decimal digits of

$$M \cdot N = e_k \cdot 10^k + e_{k-1} \cdot 10^{k-1} + \cdots + e_1 \cdot 10 + e_0$$
$$(0 \leq e_i \leq 9, e_k \geq 1)$$

we begin with $p = 0$ and compute and save the decimal digits of

$$t_p = (D_p \cdot 10^p)N = 10^p \cdot (a_0^{(p)} + a_1^{(p)} \cdot 10 + a_2^{(p)} \cdot 10^2 + \cdots + a_{r_p}^{(p)} \cdot 10^{r_p})$$

where

$$a_0^{(p)} \equiv D_p d_0 \pmod{10}$$

and, *carrying* the digits $(D_p d_{i-1} - a_{i-1}^{(p)})$, we obtain

$$a_i^{(p)} \equiv \{D_p d_i + D_p d_{i-1} - a_{i-1}^{(p)}\} \pmod{10}$$

for $i = 1, 2, \ldots, n$; and if $D_p d_n > 10$, we compute

$$a_{n+1}^{(p)} = D_p d_n - a_n^{(p)}.$$

(So r_p is either n or $n + 1$.) Then we increase p by 1 and compute the next t_p until we have $t_0, t_1, \ldots, t_m$. Then we *add* the t_p's to obtain, finally,

$$e_0 = a_0^{(0)}$$
$$e_1 = (a_1^{(0)} + a_0^{(1)}) \pmod{10}$$
$$e_2 = \{a_2^{(0)} + a_1^{(1)} + a_2^{(0)} + (a_1^{(0)} + a_0^{(1)}) - e_1\} \pmod{10}.$$

$$\cdot$$
$$\cdot$$
$$\cdot$$

The formula

$$a_i^{(p)} \equiv \{D_p d_i + D_p d_{i-1} - a_{i-1}^{(p)}\} \pmod{10} \qquad (i = 1, 2, \ldots, n)$$

is an example of a *recursion formula*. The result, $a_i^{(p)}$, of each successive evaluation of the formula is substituted into the formula for the next evaluation during the *generation* (by repeated use of the formula) of a sequence of results $a_1^{(p)}, a_2^{(p)}, \ldots, a_n^{(p)}$.

A very basic and useful recursion formula is the *nesting* method for evaluating a polynomial

$$p(x) = a_n x^n + a_{n-1} x^{n-1} + \cdots + a_1 x + a_0.$$

We can put

$$y_0 = a_n$$
$$y_i = x y_{i-1} + a_{n-i} \qquad (i = 1, 2, \ldots, n);$$

then $p(x) = y_n$.

This way of evaluating $p(x)$ requires n multiplications and additions and is both faster and has less round-off error than computing and summing the powers of x.

The Gaussian elimination method to be discussed in Chapter 4 is described by means of recursion formulas.

Numerical values of Taylor series coefficients can be generated recursively as will be shown in Sections 6.2 and 6.4.

In an *iterative method* for solving an equation, a sequence of results is computed the successive terms of which are supposed to come closer and closer to *satisfying* the equation. Suppose we wish to find a solution, x, to an equation

$$f(x) = 0.$$

Let h be some chosen nonvanishing function and put $g(y) = y + h(y)f(y)$. Then $f(x) = 0$ whenever $g(x) = x$.

A straightforward iterative method for finding x such that $f(x) = 0$ is the following:

1. Choose y_0 and put $p = 1$.
2. Compute $y_p = g(y_{p-1})$.
3. Test a criterion for stopping; if the criterion *is* met, stop with y_p as an approximation to x.
4. If the criterion *is not* met, increase p by 1, replace y_{p-1} by y_p, and return to step (2).

Notice that the equality $g(y_p) = y_p$ would imply that

$$y_p = g(y_p) = y_p + h(y_p)f(y_p)$$

and so $h(y_p)f(y_p) = 0$.

Since we have *chosen h* so that $h(y) \neq 0$, it follows that $f(y_p) = 0$. Thus if y_p is a *fixed point* of g, $(g(y_p) = y_p)$, then y_p is a *solution* of $f(x) = 0$ (a *zero* of f). Suppose f is a real-valued function of a real variable with a graph that looks, near a solution of $f(x) = 0$, like Figure 2.1.

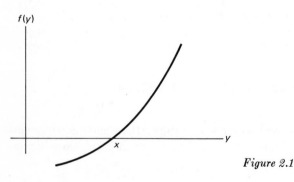

Figure 2.1

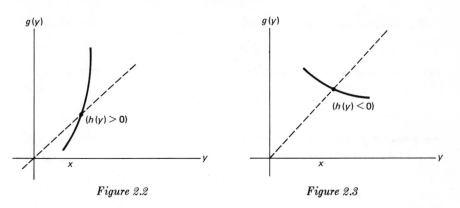

Figure 2.2   *Figure 2.3*

For what kinds of functions h will the iterative method, just discussed, work? What will a plot of $g(y) = y + h(y)f(y)$ look like? Near the solution of $f(x) = 0$ it should look something like Figure 2.2 or Figure 2.3 (the dotted lines have slope 1).

What kind of an $h(y)$ do we need so that y_p, given by

$$y_p = g(y_{p-1}) = y_{p-1} + h(y_{p-1})f(y_{p-1}),$$

is closer to x than y_{p-1} is? Surely if $h(y)$ is chosen so that, *for y near x, g(y)* crosses the diagonal line, where $g(y) = y$, *with a slope near zero*, then that will do the job if y_{p-1} is close enough to x. The sequence of *iterates* generated by one possible such function g is shown in Figure 2.4.

Consider the following example. Let

$$f(y) = y^2 - N.$$

We wish to find x such that $f(x) = 0$; that is, we wish to find a solution of

$$x^2 - N = 0.$$

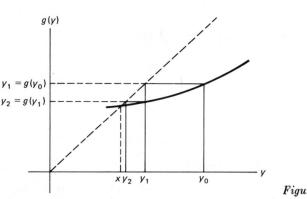

Figure 2.4

We want to choose $h(y)$ so that

$$g(y) = y + h(y)(y^2 - N)$$

has a slope near zero for y near x. Differentiating with respect to y, we obtain

$$g'(y) = 1 + h(y)(2y) + h'(y)(y^2 - N).$$

For y near x, y^2 is near $x^2 = N$, so to get $g'(y)$ near zero, we could choose $h(y)$ such that

$$1 + h(y)(2y) = 0;$$

in other words, we could choose $h(y) = -1/(2y)$. Let us *try it!* Choosing $h(y) = -1/(2y)$, we get

$$g(y) = y - \frac{y^2 - N}{2y} = \frac{1}{2}\left(y + \frac{N}{y}\right)$$

and we obtain the following iterative method for solving $f(x) = x^2 - N = 0$:

1. Choose y_0 and put $p = 1$.
2. Compute

$$y_p = \frac{1}{2}\left(y_{p-1} + \frac{N}{y_{p-1}}\right).$$

3. Test a criterion for stopping.
4. Stop or else increase p by 1; replace y_{p-1} by y_p; and return to step (2).

Leaving aside, for the moment, the criterion for stopping, let us try a few iterations of the procedure and see what happens. Let us take $N = 2$ and, for an *initial* approximation to $x^2 - 2 = 0$, take $y_0 = 1$.

The procedure yields the following sequence of *successive approximations:*

$$y_1 = \frac{1}{2}\left(1 + \frac{2}{1}\right) = \frac{3}{2} = 1.5$$

$$y_2 = \frac{1}{2}\left(1.5 + \frac{2}{1.5}\right) = \frac{17}{12} = 1.41666\ldots$$

$$y_3 = \frac{577}{408} = 1.4142156\ldots$$

.
.
.

The exact value of $x = \sqrt{2}$ has the decimal representation

$$\sqrt{2} = 1.4142135\ldots.$$

The method seems to be working well for the example we chose, at any rate.

EXERCISE

Put $N = 3$ in the iterative procedure just described for solving $x^2 - N = 0$. Set $y_0 = 2$ and compute a few iterations. Compare your result with published tables for the value of $\sqrt{3}$. The number of correct decimal digits should approximately double on each iteration.

The iterative method just discussed for solving $f(x) = 0$, based on *successive substitutions* into a formula $g(y) = y + h(y)f(y)$, is not the only type of iterative method available. Some iterative methods for solving $f(x) = 0$ begin with a *set of two or more starting approximate values* of x and repeatedly apply some procedure for adding new approximate values to this set — often discarding some of the previous values in the process.

One such procedure is the *bisection method* for solving $f(x) = 0$ when f is any continuous real-valued function with domain consisting of the real numbers in some interval $[a,b]$. To begin the bisection method we must first find (or be given) a pair of numbers a_0 and b_0 such that $a \leq a_0 < b_0 \leq b$, and such that $f(a_0)$ and $f(b_0)$ have opposite *signs*, say $f(a_0) < 0$ and $f(b_0) > 0$.

From the continuity of f it follows that at least one solution x to $f(x) = 0$ lies in the interval $[a_0, b_0]$ (see Figure 2.5). We then *bisect* the interval $[a_0, b_0]$, computing its *midpoint* as

$$m = \frac{a_0 + b_0}{2}.$$

Next we compute $f(m)$ and test its sign. If $f(m) < 0$, then we replace a_0 by $a_1 = m$ and keep $b_1 = b_0$. A solution x now lies in the narrower interval $[a_1, b_1]$. In fact, $b_1 - a_1 = \frac{1}{2}(b_0 - a_0)$, so the width of the new interval containing a solution x is half the width of the previous one.

In the other case, namely $f(m) > 0$, we replace b_0 by $b_1 = m$ and keep $a_1 = a_0$. Again we get a new interval $[a_1, b_1]$ which contains x and which is half the width of $[a_0, b_0]$. In the *very* special case that $f(m)$ should turn out exactly zero, we have $m = x$, of course, and can stop right there. This

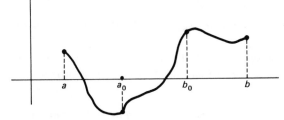

Figure 2.5

method is fairly *slowly* convergent. If $b_0 - a_0 = 1$, then after ten iterations, we will have $b_{10} - a_{10} = 2^{-10} \approx .001$.

EXERCISES

1. Apply the bisection method to the problem of finding a zero of $f(x) = x^2 - 2$ using $a_0 = 1$, $b_0 = 2$. Compute three iterations of the bisection method. Estimate the number of iterations needed by the bisection method to get $\sqrt{2}$ in this way to the accuracy which was obtained by *three* iterations of the iterative method previously discussed {which used $y_p = \frac{1}{2}[y_{p-1} + (2/y_{p-1})]$}.
2. How many iterations of the bisection method are required to get the zero of $f(x) = x^3 - 3$ in $[1,2]$ using $a_0 = 1$, $b_0 = 2$ to better than one decimal place (so error is less than 0.1)?
3. A computer subroutine takes 0.03 minute for each evaluation of a certain function $F(x)$. Computer time costs \$10 per minute. Assume that F is continuous for x in $[0,1]$ and that $F(0) < 0$ and $F(1) > 0$. How much will it cost to find a zero of F to four decimal places using the bisection method?

2.2   Convergence of Iterative Methods

In this section we consider the limiting behavior (as k gets large) of approximations given by iterative methods of the form

$$y^{(k+1)} = g(y^{(k)})$$

to solutions of equations of the form

$$f(x) = 0$$

where g is chosen to depend on f, of course.

If the $y^{(1)}$, $y^{(2)}$, . . . generated by iteratively evaluating g beginning at $y^{(0)}$ converges to a limit x, then we say that the iterative method *converges to x from $y^{(0)}$*.

A useful setting in which to study convergence of iterative methods is the concept of a *contraction mapping*.

Let S be a set of *points* (these may be real numbers or vectors) in the domain of g. We say that g is a *contraction mapping on S* if g satisfies the following two properties.

1. *g maps S into a subset of S; that is, if for every y in S, $g(y)$ is also in S.*

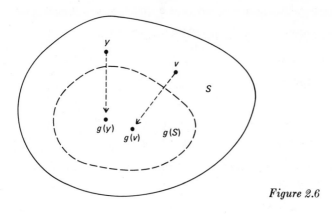

Figure 2.6

2. There is a constant c, with $0 \le c < 1$ such that, for every two points y and v in S, the *distance** $d[g(y),g(v)]$ between $g(y)$ and $g(v)$ is no greater than c times the *distance** $d(y,v)$ between y and v:

$$d[g(y),g(v)] \le cd(y,v) \qquad \text{for all } y, v \text{ in } S.$$

We can picture a contraction mapping g on S, as shown in Figure 2.6. The entire set S is shrunk or *contracted* into the smaller dotted region by the *mapping* (function) g. And each pair of points y, v in S is brought closer together by the mapping g.

We denote by $g(S)$ the *image* of S under the mapping g, that is,

$$g(S) = [g(y) \mid y \in S].$$

So $g(S)$ is the set of points into which S is mapped by g and $g(S) \subset S$. If we repeatedly apply the mapping g, we obtain a succession of smaller and smaller image sets that will contract in the limit to a point x such that $g(x) = x$, provided that the set S is *closed;* that is, it contains all its limit points. The iterative method $x^{(k+1)} = g(x^{(k)})$ will converge to x from every point $x^{(0)}$ in S.

We now give a proof of the assertions we have just made. We use the two defining properties of a contraction mapping.

Suppose g is a contraction mapping on a closed set S. Property (1) states that $g(y)$ will be in S whenever y is in S. Let $y^{(0)}$ be any point in S. Consider the sequence $y^{(1)}$, $y^{(2)}$, ... of points in S generated by $y^{(k+1)} = g(y^{(k)})$. We have

$$y^{(1)} = g(y^{(0)})$$
$$y^{(2)} = g(y^{(1)})$$

*In a vector space with a *norm* defined (see Appendix B) we can take $d(y,v) = ||y - v||$.

and property (2) states that there is a constant c with $0 \le c < 1$ such that

$$d(y^{(1)},y^{(2)}) = d[g(y^{(0)}), g(y^{(1)})] \le cd(y^{(0)},y^{(1)}).$$

Similarly, for any $p = 0, 1, \ldots$.

$$d(y^{(p+2)},y^{(p+1)}) \le cd(y^{(p+1)},y^{(p)})$$

and

$$d(y^{(p+3)},y^{(p+2)}) \le c^2 d(y^{(p+1)},y^{(p)})$$
$$d(y^{(p+k+1)},y^{(p+k)}) \le c^k d(y^{(p+1)},y^{(p)}).$$

Now we can make use of a property of distance known as the *triangle inequality*. If u, v, and y are *any* three points, then

$$d(u,v) \le d(u,y) + d(y,v).$$

This says that the length of any side of a triangle cannot exceed the sum of lengths of the other two sides (see Figure 2.7). The case of equality can occur only when the three points u, v, and y lie on a straight line.

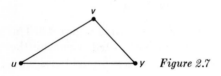

Figure 2.7

Using the triangle inequality, we can now write

$$d(y^{(2)},y^{(0)}) \le d(y^{(2)},y^{(1)}) + d(y^{(1)},y^{(0)})$$

and for any $k = 1, 2, \ldots$,

$$d(y^{(k+2)},y^{(k)}) \le d(y^{(k+2)},y^{(k+1)}) + d(y^{(k+1)},y^{(k)}).$$

Putting this last inequality together with the use of property (2) we obtain

$$d(y^{(k+2)},y^{(k)}) \le (1 + c)d(y^{(k+1)},y^{(k)}).$$

Since $d(y^{(k+3)},y^{(k+2)}) \le c^2 d(y^{(k+1)},y^{(k)})$ we have

$$d(y^{(k+3)},y^{(k)}) \le d(y^{(k+3)},y^{(k+2)}) + d(y^{(k+2)},y^{(k)})$$
$$\le (1 + c + c^2)d(y^{(k+1)},y^{(k)}).$$

It is clear (by *finite mathematical induction*) that for *any* nonnegative integers k and p we will have

$$d(y^{(k+p+1)},y^{(k)}) \le (1 + c + \cdots + c^p)d(y^{(k+1)},y^{(k)}).$$

Now using $d(y^{(p+k+1)},y^{(p+k)}) \le c^k d(y^{(p+1)},y^{(p)})$ again, this time with $p = 0$, we obtain

$$d(y^{(k+1)}, y^{(k)}) \leq c^k d(y^{(1)}, y^{(0)}).$$

Putting this together with the previous result, we obtain for any k and p, the result that

$$d(y^{(k+p+1)}, y^{(k)}) \leq (1 + c + \cdots + c^p) c^k d(y^{(1)}, y^{(0)}).$$

Since $c < 1$, the sum

$$1 + c + c^2 + \cdots + c^p = \frac{1 - c^{p+1}}{1 - c}$$

is less than $1/(1 - c)$ for any p.

We have shown that the distance between $y^{(k)}$ and $y^{(k+p+1)}$ for any p is bounded by

$$d(y^{(k+p+1)}, y^{(k)}) \leq \frac{c^k}{1 - c} \, d(y^{(1)}, y^{(0)}).$$

Now $c < 1$ so we can make c^k as small as we please by taking k large enough. Thus, for large enough k, the distance between $y^{(k)}$ and *all* the iterates beyond $y^{(k)}$, namely $y^{(k+1)}$, $y^{(k+2)}$, . . . can be made as small as we please. This means that $y^{(k)}$ converges to a limit y as $k \to \infty$. Furthermore, since

$$\lim_{k \to \infty} d(y^{(k+1)}, y^{(k)}) = \lim_{k \to \infty} d[g(y^{(k)}), y^{(k)})]$$
$$= d[(g(y), y)] = 0$$

we have $g(y) = y$.

EXERCISES

1. Show that $d(y, y^{(k+1)}) \leq cd(y, y^{(k)})$.
2. Show that the *error* in the kth iterate is bounded by

$$d(y, y^{(k)}) \leq \frac{c^k}{1 - c} \, d(y^{(1)}, y^{(0)}).$$

chapter 3

Local linear approximation

3.1 Local Linearization

During the Middle Ages, we are told, it was commonly believed that the world was flat. This was a confusion between *local* linearity and *global* linearity. We know now that the earth is more nearly a sphere (of radius about 4000 miles) than a flat disc. We have seen it from space and it is not *globally* flat. On the other hand, if we establish a *level* line of sight at some place on the earth and note some point on a distant object in our line of sight — an object much farther away than one mile — we can then move one mile *closer* to that object and again sight it. Our new line of sight to the previously noted point on the object will rise approximately 1/4000 radian from the level at our new place of observation. This is only about 0.014 degree of arc or about 51 *seconds* of arc. It is a small angle, so that we may say that within a disc of radius one mile on the earth's surface, the earth is *approximately* flat in the sense that the level lines of sight (say determined by plumb bobs) at various places within the disc will all be parallel within less than 51 seconds of arc (1/4000 radian). The earth *is* locally flat (or *linear* — stays close to a *plane*) over *small* areas to a good approximation. For a disc of radius 100 feet, the elevations of level lines of sight will vary by less than one second of arc.

The ancient Greeks knew that the ratio of the circumference to the diameter of a circle is the same number, which we call π, for *all* circles. They knew that it is also true for any circle that this number π is the ratio of the area divided by the square of the radius. Archimedes found a means of getting arbitrarily good upper and lower bounds to the number π by constructing inscribed and circumscribed polygons whose areas and circumferences he could express in terms of rational arithmetic and square roots. He discovered the local linearity of the circle and, in doing so, his work foreshadowed the eventual formalization and generalization of such methods by Newton and Leibniz into the techniques now known as *the calculus.* (Newton did not publish his major work until he was satisfied that he had written it in the style of Euclid.) The idea of *local linearization* is related to the idea of *differentiability.*

We say that a continuous function f is *differentiable* at a point x if there is a unique continuous function T_x, which satisfies the equation

$$f(y) = f(x) + T_x(y)(y - x).$$

We call $T_x(x) = f'(x)$, the *derivative of f at x.*

For instance, if f is defined on the real line (the domain of f) by $f(y) = y^2$, then the equation $f(y) = f(x) + T_x(y)(y - x)$ becomes

$$y^2 = x^2 + T_x(y)(y - x)$$

and has the unique solution $T_x(y) = y + x$. The derivative of f at x is $f'(x) = T_x(x) = 2x$.

If f is defined on the complex plane and $f(z) = z^2$, then for any complex numbers z_1 and z_2, the equation

$$z_2{}^2 = z_1{}^2 + T_{z_1}(z_2)(z_2 - z_1)$$

has the unique solution $T_{z_1}(z_2) = z_2 + z_1$ and so $f'(z_1) = 2z_1$. If f is a linear transformation* on E^n defined by

$$f(x) = Ax$$

where x is a point in E^n represented as a column vector and A is an $n \times n$ matrix, then the equation

$$f(x) = f(x^{(0)}) + T_{x^{(0)}}(x)(x - x^{(0)})$$

becomes

$$Ax = Ax^{(0)} + T_{x^{(0)}}(x)(x - x^{(0)})$$

and has the unique solution

$$T_{x^{(0)}}(x) = A$$

and so

$$f'(x) = T_x(x) = A.$$

*See Appendix B.

Notice that the *derivative* of a linear transformation is the *same* linear transformation.

As a final illustration, let f be a nonlinear transformation of E^n given by a system of n nonlinear formulas for the components of $f(x)$ as

$$f(x) = \begin{pmatrix} f_1(x_1, & \ldots, & x_n) \\ f_2(x_1, & \ldots, & x_n) \\ \cdot & & \\ \cdot & & \\ \cdot & & \\ f_n(x_1, & \ldots, & x_n) \end{pmatrix}$$

where

$$x = \begin{pmatrix} x_1 \\ x_2 \\ \cdot \\ \cdot \\ \cdot \\ x_n \end{pmatrix}$$

The equation

$$f(x) = f(x^{(0)}) + T_{x^{(0)}}(x)(x - x^{(0)}),$$

if it has a unique solution $T_{x^{(0)}}(x)$, will give the *derivative* $f'(x^{(0)}) = T_{x^{(0)}}(x^{(0)})$ as the *Jacobian matrix* of partial derivatives evaluated at $x^{(0)}$ of the system of component functions of f.

The i, jth component of the Jacobian matrix $f'(x^{(0)})$ is given by

$$[f'(x^{(0)})]_{i,j} = \frac{\partial f_i(x_i^{(0)}, \ldots, x_n^{(0)})}{\partial x_j} \qquad \text{for } i, j = 1, 2, \ldots, n.$$

For example, if f is given as a nonlinear transformation of the plane (E^2) by

$$f_1(x_1,x_2) = x_1{}^2 + x_2{}^2$$
$$f_2(x_1,x_2) = x_1 - x_2$$

then

$$f'(x^{(0)}) = \begin{pmatrix} 2x_1^{(0)} & 2x_2^{(0)} \\ 1 & -1 \end{pmatrix}.$$

EXERCISES

Find the derivative of each of the following:

1. $f(x) = \begin{pmatrix} 0 & 1 \\ -1 & 0 \end{pmatrix}\begin{pmatrix} x_1 \\ x_2 \end{pmatrix}, \; x = \begin{pmatrix} x_1 \\ x_2 \end{pmatrix}.$

2. $f(x) = e^x - x.$

3. $f(x) = \begin{pmatrix} f_1(x_1,x_2) \\ f_2(x_1,x_2) \end{pmatrix},$ where

$$f_1(x_1,x_2) = x_1 - 2x_2 + x_1 x_2$$
$$f_2(x_1,x_2) = x_1^2 + x_2^2 - x_1.$$

With these definitions and illustrations of derivatives of differentiable functions in mind, let f be a differentiable function; then there is a unique function T such that

$$f(y) = f(x) + T_x(y)(y - x)$$

and the derivative of f at x is

$$f'(x) = T_x(x).$$

If we hold x fixed, then the expression

$$f(x) + f'(x)(y - x)$$

is *linear* in y and gives a *linear approximation* to $f(y)$.

If $T_x(y)$ is continuous, then for y *near* x, $T_x(y)$ will be *near* $T_x(x) = f'(x)$ and the linear approximation $f(x) + f'(x)(y - x)$ will be near $f(y)$. In fact, the accuracy of the linear approximation is expressed by the equation

$$f(y) - [f(x) + f'(x)(y - x)] = [T_x(y) - T_x(x)](y - x).$$

If we use $f(x) + f'(x)(y - x)$ in place of $f(y)$ (for whatever purpose), this replacement is called the *local linearization of f at x*. The approximation is *linear* in y and is *local*, that is, it is a good approximation for *y near x*.

Suppose f is a real-valued differentiable function of a single real variable with a graph perhaps as shown in Figure 3.1. The derivative of f at x is the *slope* of the tangent line to the graph of f at the point $[x, f(x)]$. For y near x the graph of f is near the tangent line

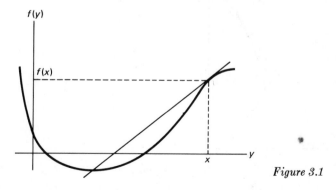

Figure 3.1

$$L(y) = f(x) + f'(x)(y - x).$$

In fact,

$$f(y) - L(y) = f(y) - f(x) - f'(x)(y - x)$$

and for $y > x$, say, we have

$$f(y) - L(y) = \left[\frac{f(y) - f(x)}{y - x} - f'(x)\right](y - x)$$

and the quantity in brackets is the difference between the slopes of the following:

1. The line through the two points $[x, f(x)]$, $[y, f(y)]$.
2. The tangent line to the graph of f at x, as shown in Figure 3.2. If we keep x fixed and let y get closer to x, then the slopes will get closer.

EXERCISE

If the point (x_1, x_2) lies on the unit circle, then

$$x_1^2 + x_2^2 = 1.$$

For points on the upper half circle we can put

$$x_2 = \sqrt{1 - x_1^2}.$$

Find the local linearization of

$$x_2(x_1) = \sqrt{1 - x_1^2}$$

at $x_1 = \frac{1}{2}$.

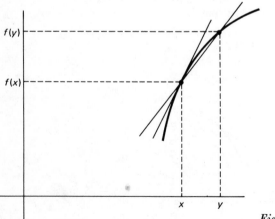

Figure 3.2

3.2 Newton's Method

A commonly used and often very efficient iterative method for solving non-linear equations is *Newton's method*. Given an initial guess $y^{(0)}$ to the solution of an equation $f(x) = 0$, we replace f by its local linearization near $y^{(0)}$, namely

$$L(y) = f(y^{(0)}) + f'(y^{(0)})(y - y^{(0)})$$

and *solve* the linear equation

$$L(y) = 0$$

to get a (hopefully) better approximation than $y^{(0)}$ to the solution of $f(x) = 0$. We can repeat (iterate) this process.

Figure 3.3 illustrates graphically the first two successive iterates obtained by this method. The iterates in Newton's method are obtained by solving

$$f'(y^{(k)})(y^{(k+1)} - y^{(k)}) = -f(y^{(k)})$$

successively for $y^{(1)}$, $y^{(2)}, \ldots$, beginning with a given starting guess $y^{(0)}$.

To get an explicit iteration formula, we can write

$$y^{(k+1)} = y^{(k)} - [f'(y^{(k)})]^{-1}f(y^{(k)}).$$

In this form, Newton's method can be viewed as the iteration method we get for solving $f(x) = 0$ with the iteration function

$$g(y) = y + h(y)f(y)$$

when $h(y)$ is chosen as $-[f'(y)]^{-1}$.

The iterative method discussed in Chapter 2 for solving $f(x) = x^2 - N = 0$ was, in fact, Newton's method. We had

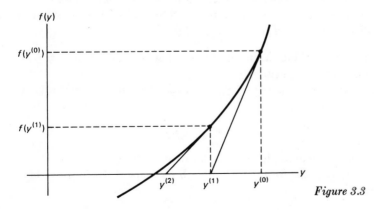

Figure 3.3

$$y_p = \frac{1}{2}\left(y_{p-1} + \frac{N}{y_{p-1}}\right);$$

and for $N = 2$ and $y_0 = 1$ we obtained

$$y_1 = 1.5$$
$$y_2 = 1.416\ldots$$
$$y_3 = 1.4142156\ldots$$

$$\cdot$$
$$\cdot$$
$$\cdot$$

For systems of equations, when f' is a matrix, the form $f'(y^{(k)})(y^{(k+1)} - y^{(k)})$ $= -f(y^{(k)})$ is more efficient. We can solve the linear system for $y^{(k+1)} - y^{(k)}$ instead of trying to invert the matrix (see Chapter 4).

EXERCISES

1.   Express the iteration formula given by Newton's method for solving $f(x) = 0$ when
     (a) $f(y) = y^2 - N$.
     (b) $f(y) = a_n y^n + a_{n-1} y^{n-1} + \cdots + a_1 y + a_0$.
     (c) $f(y) = y - e^{-y}$.
     (d) $f(y) = \begin{pmatrix} y_1^2 + y_2^2 - 1 \\ y_1 - y_2 \end{pmatrix}$ with $y = \begin{pmatrix} y_1 \\ y_2 \end{pmatrix}$.
     (e) $f(z) = z^2 + 2$ for complex $z = (z_1, z_2)$.
2.   Find suitable initial guesses for Newton's method as applied to (c), (d), and (e) in Exercise 1 and use the method to compute approximate solutions accurate to three decimal places.

If $g(y)$ is a *real-valued* function of the form $g(y) = y + h(y)f(y)$ for real-valued continuous functions h and f, then the conditions that g must satisfy in order to be a contraction on an interval $S = [a,b]$ of values of y are [*distance* in this case is just $d(y,v) = |y - v|$]:

1. g maps $S = [a,b]$ into a subset of $[a,b]$; that is, for every $a \le y \le b$, we must have $a \le g(y) = y + h(y)f(y) \le b$.
2.   There is a constant c, with $0 \le c < 1$ such that for every y and v in $[a,b]$, we have

$$|g(y) - g(v)| \le c|y - v|.$$

Let us examine these conditions in the case of Newton's method for solving an equation of the form $f(x) = 0$ with f a real-valued function of a single real variable x.

For Newton's method, we have $h(y) = -[f'(y)]^{-1}$ and so condition (1) becomes:

1. $a \leq g(y) = y - [f'(y)]^{-1}f(y) \leq b$ for every y in $[a,b]$.
2. In the second condition there is a constant c, $0 \leq c < 1$ such that for every y and v in $[a,b]$ we have

$$|y - [f'(y)]^{-1}f(y) - \{v - [f'(v)]^{-1}f(v)\}| \leq c|y - v|.$$

If $0 < \alpha \leq f'(y) \leq \beta$ for all y in $[a,b]$, then

$$\min\left(a - \frac{f(a)}{\beta}, b - \frac{f(b)}{\alpha}\right) \leq y - [f'(y)]^{-1}f(y) \leq \max\left(a - \frac{f(a)}{\alpha}, b - \frac{f(b)}{\beta}\right).$$

See Figure 3.4 and also Figure 3.3. Thus, to satisfy condition (1) we need $[a,b]$ such that

3. $0 < \alpha \leq f'(y) \leq \beta$ for y in $[a,b]$.
4. $a \leq \min\left(a - \frac{f(a)}{\beta}, b - \frac{f(b)}{\alpha}\right)$.
5. $b \geq \max\left(a - \frac{f(a)}{\alpha}, b - \frac{f(b)}{\beta}\right)$.

These conditions imply, in particular, that we must have $f(a) \leq 0$ and $f(b) \geq 0$. If we had an f for which $f'(y) < 0$ for y in $[a,b]$, we could treat the function $(-f)$ by the method under discussion.

Turning now to condition (2), we can write (if f is twice continuously differentiable)

$$f(y) = f(v) + f'(v)(y - v) + \frac{f''(\xi_1)}{2}(y - v)^2$$

and

$$f(v) = f(y) + f'(y)(v - y) + \frac{f''(\xi_2)}{2}(v - y)^2$$

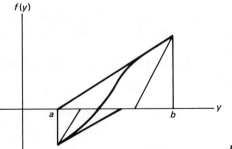

Figure 3.4

for some ξ_1, ξ_2 between v and y; and we can use these expressions to rewrite condition (2) as follows:

$$|y - (f'(y))^{-1}f(y) - \{v - [f'(v)]^{-1}f(v)\}|$$
$$= |(y - v) - [f'(v)]^{-1}[f(y) - f(v)] + \{[f'(v)]^{-1} - [f'(y)]^{-1}\}f(y)|$$
$$= |- [f'(v)]^{-1}\frac{f''(\xi_1)}{2}(y - v)^2 - \frac{f''(\xi_1) + f''(\xi_2)}{2f'(v)f'(y)}f(y)(v - y)|$$
$$\leq c|y - v| \qquad \text{for all } y, v \text{ in } [a,b].$$

Suppose, as in condition (3), that $0 < \alpha \leq f'(y) \leq \beta$ for y in $[a,b]$ and suppose further that we have

6. $|f''(\xi)| \leq \gamma$ for all ξ in $[a,b]$; then we can satisfy condition (2) if we can find a constant c such that

7. $\dfrac{\gamma}{2\alpha}|b - a| + \dfrac{\gamma}{\alpha^2}|f(y)| \leq c < 1$ for every y in $[a,b]$.

Since we are supposing [condition (3)] that $f'(y)$ is positive, the function $f(y)$ must be monotonic increasing, so we could replace (7) by

8. $\dfrac{\gamma}{2\alpha}|b - a| + \dfrac{\gamma}{\alpha^2}\max [|f(a)|,f(b)] \leq c < 1.$

If conditions (3), (4), (5), (6), and (8) are satisfied, then the function

$$g(y) = y - [f'(y)]^{-1}f(y)$$

is a contraction mapping on the interval $[a,b]$ and we may take *any* $x^{(0)}$ in $[a,b]$ as a starting point and Newton's iterative method will converge to a zero of f in $[a,b]$.

EXERCISE

Show that all the conditions (3), (4), (5), (6), and (8) are satisfied for y in the interval [1,1.7] for the function $f(y) = y^2 - 2$; so that Newton's method applied to $f(y) = y^2 - 2 = 0$ will converge from any $x^{(0)}$ in [1,1.7].

An *interval version of Newton's method* was introduced in Moore, 1962 (see also Moore, 1966; Nickel, 1969; Nickel, 1971). It is based on the mean-value theorem and on interval contractions.

If f has a continuous derivative in $[a,b]$, then for x and y in $[a,b]$ we have

$$f(x) = f(y) + f'(y + \theta(x - y))(x - y)$$

for some θ in [0,1]. Assume that $f(x) = 0$ for some x in $[a,b]$. Thus

$$x = y - \frac{f(y)}{f'(y + \theta(x - y))}$$

Denote the midpoint* of an interval $X = [a,b]$ by $m(X) = (a + b)/2$. Define the interval function N by

$$N(X) = m(X) - \frac{f[m(X)]}{f'(X)},$$

where $f'(X)$ contains $f'(x)$ for every x in X; then the interval version of Newton's method (for one dimension) is as follows:

$$X_{n+1} = X_n \cap N(X_n).$$

If the intersection is empty, there is no zero of f in X_n.

It is shown (Moore, 1962 and 1969) that $x \in X_0$ implies $x \in X_n$ for all $n = 1, 2, \ldots$. As an illustration, let $f(x) = x^2 - 2$, with $X_0 = [1,2]$. Now $\sqrt{2}$ is in $[1,2]$, therefore it is also in X_n for every n, where

$$X_{n+1} = X_n \cap \left\{ m(X_n) - \frac{[m(X_n)]^2 - 2}{2X_n} \right\}$$

is computed by the interval Newton method. We obtain, for instance,

$$N(X_0) = \left[\frac{22}{16}, \frac{23}{16} \right] \subset X_0$$

so that, in fact, we have

$$X_1 = \left[\frac{22}{16}, \frac{23}{16} \right] = X_0 \cap N(X_0)$$
$$X_2 = [1.41406\ldots, 1.41441\ldots]$$
$$X_3 = [1.414213559\ldots, 1.414213566\ldots]$$

.

.

.

If we can find an interval X_0 such that $f'(X_0)$ does not contain zero and such that f is opposite sign at the left end-point and right end-point of X_0 [which is easier to do than to find an interval $[a,b]$ that satisfies conditions (3), (4), (5), (6), and (8)], then each X_n will also contain the zero of f in X_0, and the intervals $X_1, X_2, \ldots$ will shrink toward a narrow interval of width, depending on the *precision* of rounded interval arithmetic which is used.

The following example will further illustrate some differences between the ordinary and interval versions of Newton's method. Consider the polynomial

$$f(x) = x^3 - 2x^2 - x + 1.$$

We find that

*Actually the method will allow *any* point in X for $m(X)$.

$$f(-1) = -1, \quad f(0) = 1, \quad f(\tfrac{3}{2}) = -\frac{13}{8}, \quad f(3) = 7.$$

From the sign changes in the values of f we can see that this cubic has three real zeros: one in the interval $[-1,0]$, one in $[0,\tfrac{3}{2}]$, and one in $[\tfrac{3}{2},3]$.

Let us try to find the middle zero (the one in $[0,\tfrac{3}{2}]$) using the two versions of Newton's method. Now $f'(x) = 3x^2 - 4x - 1$ and we find that $f'(\tfrac{3}{2}) = -\tfrac{1}{4}$. If we choose the point $x_0 = \tfrac{3}{2}$ as a starting value in the ordinary Newton method, we obtain

$$x_1 = x_0 - \frac{f(x_0)}{f'(x_0)} = -5.9\cdots$$

and from this point the successive values $x_2, x_3, \ldots$ obtained from Newton's method will converge to the zero in $[-1,0]$ instead of the one we want in $[0,\tfrac{3}{2}]$. However, if we put

$$X_0 = [0,\tfrac{3}{2}]$$

in the interval version of Newton's method, we find that, using

$$\begin{aligned}
f'(X) &= X(3X - 4) - 1, \\
f'(X_0) &= [0,\tfrac{3}{2}]([0,\tfrac{9}{2}] - 4) - 1 \\
&= [0,\tfrac{3}{2}][-4,\tfrac{1}{2}] - 1 \\
&= [-6,\tfrac{3}{4}] - 1 \\
&= [-7,-\tfrac{1}{4}].
\end{aligned}$$

Thus $m(X_0) = \tfrac{3}{4}, f[m(X_0)] = -29/64$

$$\begin{aligned}
N(X_0) &= \frac{3}{4} - \frac{29/64}{[-7,-\tfrac{1}{4}]} \\
&= \left[-\frac{17}{16}, \frac{307}{448}\right]
\end{aligned}$$

Next, we compute the intersection

$$\begin{aligned}
X_1 &= X_0 \cap N(X_0) \\
&= \left[0,\frac{3}{2}\right] \cap \left[-\frac{17}{16}, \frac{307}{448}\right] \\
&= \left[0,\frac{307}{448}\right].
\end{aligned}$$

We now know that the middle zero of f is in the narrower interval X_1. From X_1 we can compute

$$X_2 = X_1 \cap N(X_1).$$

We obtain

$$N(X_1) = N\left(\left[0, \frac{307}{448}\right]\right) = [.46\ldots, .8\ldots]$$

and so

$$X_2 = [0, .68\ldots] \cap [.46\ldots, .8\ldots] = [.46\ldots, .68\ldots].$$

From this point on the successive intervals will *shrink* in width very rapidly; the widths will be roughly *squared* each time, until the limit of machine precision is reached.

EXERCISE

Program and carry out the computation of X_1, X_2, X_3, X_4, X_5 for the preceding example using rounded interval arithmetic on the computer.

The verification of conditions for a contraction mapping in the case of Newton's method or some other iterative method applied to the solution of a system of nonlinear equations is necessarily more complicated than the scalar case which we have carried out, and therefore, we will omit it; but it should be noted, at least, that the conclusions of the contraction mapping principle will apply if a region can be found for which the conditions are satisfied.

It is possible to combine interval arithmetic with Newton's method to obtain rigorous error bounds on numerical solutions of n-dimensional *systems* of nonlinear equations (including zeros of polynomials as a special one-dimensional case) (Hansen, 1969; Kuba and Rall, 1972; Moore, 1962 and 1966; Nickel, 1971).

In the practical use of iterative methods for finding solutions of equations of the form $f(x) = 0$, a reasonable criterion for stopping an iteration is to test the value of f (or the value of a norm of f if f is a vector-valued function). If we find an $x^{(k)}$ such that $|f(x^{(k)})|$ (or $\|f(x^{(k)})\|$) is sufficiently *small* according to some reasonable practical criterion, then we may have a satisfactory approximation to a solution of the equation. As pointed out in Section 3.3, if $\bar{x}$ is regarded as an approximation to an x such that $f(x) = 0$, we can estimate $x - \bar{x}$ as

$$x - \bar{x} \approx -(f'(\bar{x}))^{-1}f(\bar{x})$$

based on the locally linear approximation

$$f(x) \approx f(\bar{x}) + f'(\bar{x})(x - \bar{x}).$$

For vector equations $f(x) = 0$, we can use the same technique with $f'(\bar{x})$ as a *matrix* in this case.

Suppose we carry out a couple of iterations of Newton's method on the equation

$$f(x) = x^2 - 2 = 0$$

starting with $y^{(0)} = 1$. We find that $f'(x) = 2x$, so that

$$y^{(k+1)} = y^{(k)} - [f'(y^{(k)})]^{-1} f(y^{(k)}) = y^{(k)} - \frac{1}{2y^{(k)}} [(y^{(k)})^2 - 2]$$

and

$$y^{(1)} = 1 - \tfrac{1}{2}(1 - 2) = 1.5$$
$$y^{(2)} = 1.5 - \tfrac{1}{3}(2.25 - 2) = 1.4166 \ldots.$$

We can compute (using $y^{(2)}$ in place of $\bar{y}$) $f(y^{(2)}) = .0068 \ldots$ and $[f'(y^{(2)})]^{-1} = .353 \ldots$, so we have the estimate $x - y^{(2)} \approx -.00236 \ldots.$

EXERCISES

1. Derive the *error-squaring* property of Newton's method [for $f(x) = 0$]:

$$y^{(k+1)} - x = [f'(y^{(k)})]^{-1} \frac{f''(\xi)}{2} (y^{(k)} - x)^2.$$

2. Show that Newton's method converges to a solution x of $f(x) = 0$ *from any* $y^{(0)} > x$, if f is convex to the right of x, that is, if $f''(y) > 0$ for $y > x$. Assume here that $f(y) > 0$ for $y > x$ and $f(y) < 0$ for $y < x$.

3. Show that any zero of the polynomial $x^n + a_{n-1}x^{n-1} + \cdots + a_0$ satisfies

$$|x| \le \max \{1, \sum_{i=0}^{n-1} |a_i|\}.$$

It follows from the *error-squaring* property of Newton's method that if we are close enough to a *simple* (not multiple) root of a function f so that f' and f'' do *not* vary much within that distance to the root, then the number of correct decimal digits in the iterates will roughly *double* in each iteration so that a few iterations will suffice for great accuracy.

3.3   Perturbations, Variations, and Sensitivity

Suppose we have found, by one means or another, an approximate solution $\bar{x}$ to an equation $f(x) = 0$ in which there appear certain *constants* or *parameters*, $a_0, a_1, a_2, \ldots, a_n$. We can ask what happens to a zero of f when we change one or more of the parameters $a_0, a_1, \ldots, a_n$. Put another way, what can we add to $\bar{x}$ to get an approximate solution to $f(x) = 0$ with a slightly different value for one or more of the parameters $a_0, a_1, \ldots, a_n$?

If the function f depends on the parameters in question in such a way that it is also a *differentiable function of the parameters*, then we can attempt to answer such questions in the following way. We can denote a value of f for a given value of x and a given set of values of the parameters $a_0, a_1, \ldots, a_n$ by $f(x, a_0, a_1, \ldots, a_n)$. If

$$f(x, a_0, a_1, \ldots, a_n) = 0,$$

and if $f(\bar{x}, a_0, a_1, \ldots, a_n)$ is approximately zero, then for a small real number δ,

$$f(y, a_0, a_1, \ldots, a_i + \delta, \ldots, a_n)$$

is approximately (by local linearization)

$$f(x, a_0, a_1, \ldots, a_i, \ldots, a_n) + \frac{\partial f}{\partial a_i}(x, a_0, a_1, \ldots, a_n)\delta$$

$$+ \frac{\partial f}{\partial x}(x, a_0, a_1, \ldots, a_n)(y - x).$$

If we now look for an approximate solution $\bar{y}$ of the equation

$$f(y, a_0, a_1, \ldots, a_i + \delta, \ldots, a_n) = 0,$$

we can take [using $f(\bar{x}, a_0, a_1, \ldots, a_n) \approx 0$]

$$\bar{y} = \bar{x} - \frac{(\partial f/\partial a_i)(\bar{x}, a_0, a_1, \ldots, a_n)}{(\partial f/\partial x)(\bar{x}, a_0, a_1, \ldots, a_n)}\delta.$$

This equation expresses the linear or *first-order variation* in the solution x resulting from a *perturbation* of one of the parameters a_i in the equation $f(x) = 0$. For example, if we have $\bar{x}$ as an approximate solution of $f(x, a_0, a_1, \ldots, a_n) = a_0 + a_1 x + \cdots + a_n x^n = 0$, we can take

$$\bar{y} = \bar{x} - \frac{\bar{x}^i}{(a_1 + 2a_2\bar{x} + \cdots + na_n\bar{x}^{n-1})}\delta$$

as an approximate solution of

$$a_0 + a_1 y + \cdots + (a_i + \delta)y^i + \cdots + a_n y^n = 0.$$

Let us try this on the equation

$$x^2 - 2 = 0.$$

If we take $\bar{x} = 1.4$ as an approximate solution, we can ask for an approximate solution $\bar{y}$ to the equation

$$y^2 - 2 + \delta = 0.$$

Using the notation described, we have (with $a_i = a_0 = -2$ and $a_i + \delta = a_0 + \delta = -2 + \delta$),

$$\bar{y} = \bar{x} - \frac{(\bar{x})^0}{a_1 + 2a_2\bar{x}}\, \delta$$

$$= \bar{x} - \frac{1}{2\bar{x}}\, \delta$$

$$= 1.4 - \frac{1}{2.8}\, \delta$$

$$= 1.4 - (.357\ldots)\delta.$$

Let us plot a curve showing $\bar{y}$ as a linear function of δ, as in Figure 3.5. We also plot the curve of exact solutions $y(\delta) = \sqrt{2 - \delta}$. From the figure, it is clear that we can get a pretty good approximation to the perturbed equation by linear approximation up to say $\delta = \frac{1}{2}$ or so. Beyond that, things get progressively worse.

In the example just discussed, we could say that the solution is not very *sensitive* to very small perturbations in the constant term. If we simultaneously perturb more than one parameter value, then we can use the more general expression

$$\bar{y} = \bar{x} - \frac{\displaystyle\sum_{i=0}^{n} \frac{\partial f}{\partial a_i}(\bar{x}, a_0, a_1, \ldots, a_n)\delta_i}{\dfrac{\partial f}{\partial x}(\bar{x}, a_0, a_1, \ldots, a_n)}$$

as an approximation to the solution of

$$f(y, a_0 + \delta_0, a_1 + \delta_1, \ldots, a_n + \delta_n) = 0$$

in terms of an approximate solution $\bar{x}$ to

$$f(x, a_0, a_1, \ldots, a_n) = 0.$$

For example, $\bar{x} = .99$ is an approximate solution to $2 - 3x + x^2 = 0$.

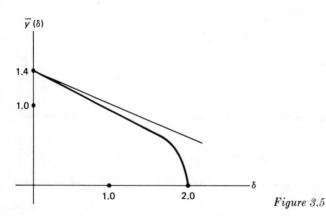

Figure 3.5

We have $f(x,2,-3,1) = 2 - 3x + x^2$, putting $f(x,a_0,a_1,a_2) = a_0 + a_1x + a_2x^2$, and the expression for $\bar{y}$ becomes, in this case,

$$\bar{y} = .99 - \frac{\delta_0 + (.99)\delta_1 + (.99)^2\delta_2}{-3 + 2(.99)}.$$

From this we could estimate that for $\delta_0 = .01$, $\delta_1 = .05$, $\delta_2 = .1$ we have

$$\bar{y} = .99 - \frac{.01 + (.99)(.05) + (.99)^2(.1)}{-3 + 2(.99)} = 1.14\ldots$$

is an approximate solution to

$$2.01 - 2.95y + 1.1y^2 = 0.$$

In fact,

$$f(\bar{y},2.01,-2.95,1.1) = f(1.14,2.01,-2.95,1.1) = .075\ldots.$$

We could try to improve on the approximation $\bar{y} = 1.14$ by applying Newton's method. To do this, put (for simplicity)

$$f(y) = 2.01 - 2.95y + 1.1y^2;$$

then

$$f'(y) = -2.95 + 2.2y$$

and Newton's method gives the following iteration formula for improving an approximate solution $y^{(k)}$ of $f(y) = 0$:

$$y^{(k+1)} = y^{(k)} - \frac{2.01 - 2.95y^{(k)} + 1.1(y^{(k)})^2}{-2.95 + 2.2y^{(k)}}.$$

We can put $y^{(0)} = 1.14$ and compute

$$y^{(1)} = 1.14 - \frac{2.01 - 2.95(1.14) + 1.1(1.14)^2}{-2.95 + 2.2(1.14)} = 1.31\ldots$$

and iterating further, we get

$$y^{(2)} = 1.79\ldots$$

and

$$y^{(3)} = 1.79 - \frac{2.01 - 2.95(1.79) + 1.1(1.79)^2}{-2.95 + 2.2(1.79)} = 1.53\ldots.$$

This seems strange; we may pause at this point to see what is going on. A diagram will help. The graph of the polynomial function

$$f(x,2,-3,1) = 2 - 3x + x^2$$

is shown in Figure 3.6.

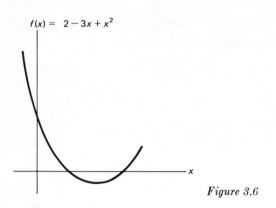

Figure 3.6

What has happened is that the changes in coefficients that we have as-sumed give the polynomial function

$$f(y, 2.01, -2.95, 1.1) = 2.01 - 2.95y + 1.1y^2$$

which has a graph as shown in Figure 3.7. This function is *positive* for all y and has *no* real zeros.

The "perturbations" of the coefficients which we made were really *not* "small" at all and together they exceeded the "range" over which the local linearization used gives an accurate approximation to the resulting change in the zero of f. It is not an easy matter to estimate this range in advance. We can at least be aware of some of the ways in which things can go wrong.

In the same example, if we take smaller perturbations in the coefficients, the techniques discussed will work.

EXERCISES

1. Estimate the change in the zero $x = 1$ of $2 - 3x + x^2$ resulting from changing the coefficients so that we have instead the polynomial

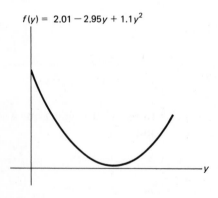

Figure 3.7

$2.01 - 2.99x + 1.02x^2$. Improve the estimate by applying Newton's method.

2. The *complex* zeros of $2.01 - 2.95y + 1.1y^2$ *can* be found to good approximation using Newton's method; but, to do it, we must start with initial approximations $z^{(0)}$ in the complex plane *off* the real line. Try $z^{(0)} = (1,1)$, for instance, and also $z^{(0)} = (1,-1)$.

Suppose we obtain, by whatever method, an approximate solution $\bar{x}$ to an equation $f(x) = 0$. How can we estimate the accuracy of $\bar{x}$? If we evaluate f at $\bar{x}$, we will have $f(\bar{x})$. If the *evaluation were exact* and $f(\bar{x}) = 0$, then $\bar{x}$ is perfectly accurate. Otherwise, we can estimate the difference between $\bar{x}$ and an unknown exact solution x by using a linear approximation to f near $\bar{x}$. If we approximate $f(y)$ for y near $\bar{x}$ by

$$f(y) \approx f(\bar{x}) + f'(\bar{x})(y - \bar{x}),$$

then in order to have x such that $f(x) = 0$ we can estimate that

$$x - \bar{x} \approx -[f'(\bar{x})]^{-1}f(\bar{x}).$$

If, in fact, there *is* a zero of f sufficiently near $\bar{x}$, and *if* f is differentiable, the estimate *may* be reasonably good.

chapter

Numerical methods in linear algebra

4.1 Introduction

Here we will consider numerical methods for solving the following two types of problems in linear algebra. (Some readers may want to review Appendix B first.)

1. Given an $n \times n$ matrix A with real coefficients A_{ij} and an n-dimensional vector written in column form as b, solve the system of linear algebraic equations for x:

$$Ax = b.$$

2. Given A as in (1), find the inverse matrix, A^{-1}, such that

$$AA^{-1} = I, \quad I_{ij} = \begin{pmatrix} 1 & \text{if } i = j \\ 0 & \text{if } i \neq j \end{pmatrix}$$

We will consider mainly two basic methods as applied to each of these problems: (1) a *direct* method giving exact results in a finite number of arithmetic operations, called *Gaussian elimination*, and (2) an *indirect* iterative method.

4.2   Gaussian Elimination for $Ax = b$

We begin with an example. Consider the linear algebraic system of equations

$$2x_1 - 3x_2 = 1$$
$$-x_1 + x_2 = 4.$$

If we multiply the second equation by 2, we clearly do not disturb any solutions the system may have and we have the equivalent system

$$2x_1 - 3x_2 = 1$$
$$-2x_1 + 2x_2 = 8.$$

If we add the first equation to the second, we obtain

$$-x_2 = 9$$

or

$$x_2 = -9.$$

If we substitute $x_2 = -9$ into the first equation, we obtain

$$2x_1 - 3(-9) = 1$$

or

$$2x_1 = 1 - 27 = -26$$

or

$$x_1 = -13.$$

We have computed the solution

$$(x_1,x_2) = (-13,-9) \quad \text{or} \quad \begin{pmatrix} x_1 \\ x_2 \end{pmatrix} = \begin{pmatrix} -13 \\ -9 \end{pmatrix}$$

in column notation. We can check that

$$2(-13) - 3(-9) = 1$$

and

$$-(-13) + (-9) = 4$$

so that $(x_1,x_2) = (-13,-9)$ does satisfy the original system of equations.

This is an example of the use of the *Gaussian elimination* method. We *eliminated* the variable x_1 by combining equations to solve for x_2. Then we used that to eliminate x_2 in another equation and found x_1.

EXERCISE

Solve the systems

1.
$$9x_1 - 8x_2 = 11$$
$$10x_1 - 9x_2 = -10$$

2.
$$x_1 + 2x_2 = 2$$
$$3x_1 + 4x_2 = -1$$
3.
$$3x_1 - 3x_2 = 0$$
$$3x_1 + 3x_2 = 0$$
4.
$$14x_1 + 7x_2 = 0$$
$$7x_1 + 14x_2 = 1$$
5.
$$2x_1 + x_2 = 1$$
$$4x_1 + 2x_2 = 3$$
6.
$$2x_1 + x_2 = 1$$
$$4x_1 + 2x_2 = 2$$

Next, we will discuss the application of the Gaussian elimination method to the solution of any system of *three* linear algebraic equations in three unknowns x_1, x_2, x_3. In this case, the system has the form

$$Ax = b$$

where

$$A = \begin{pmatrix} A_{11} & A_{12} & A_{13} \\ A_{21} & A_{22} & A_{23} \\ A_{31} & A_{32} & A_{33} \end{pmatrix}$$

and

$$b = \begin{pmatrix} b_1 \\ b_2 \\ b_3 \end{pmatrix}.$$

We seek

$$x = \begin{pmatrix} x_1 \\ x_2 \\ x_3 \end{pmatrix}$$

such that

(1) $$A_{11}x_1 + A_{12}x_2 + A_{13}x_3 = b_1,$$
(2) $$A_{21}x_1 + A_{22}x_2 + A_{23}x_3 = b_2,$$
(3) $$A_{31}x_1 + A_{32}x_2 + A_{33}x_3 = b_3.$$

Clearly, if we write these three equations in some other order, for instance,

(2)   (3)   (1)
(3)   or   (2)   or   (3)   . . .
(1)   (1)   (2)

we will have an equivalent system of equations, that is, the same x_1, x_2, x_3 (if any) will solve the equations no matter in what order the equations are written.

Each of the three equations (1), (2), and (3) is the equation of a *plane* in three-dimensional vector space E^3. Two of the equations taken together, for instance, (1) *and* (2), or (1) *and* (3), or (2) *and* (3), represent a pair of planes. If (x_1,x_2,x_3) is a three-tuple, (three-dimensional vector), representing the coordinates of a point x in E^3, then (x_1,x_2,x_3) will satisfy the first equation if and only if the point x lies in the plane which that equation represents. Furthermore, x will satisfy a pair of the equations, say (1) and (2), if and only if x lies in *both* of the planes, that is, if x lies in the *intersection* of the two planes.

Now two planes in E^3 will *intersect* in a *line* unless the planes are parallel. The condition for (1) and (2) to represent parallel planes is that the row vector (A_{11},A_{12},A_{13}) is a scalar multiple of (A_{21},A_{22},A_{23}), say $(A_{11},A_{12},A_{13}) = a(A_{21},A_{22},A_{23})$. (This is the same as saying that for some a, $A_{11} = aA_{21}$, $A_{12} = aA_{22}$, $A_{13} = aA_{23}$.)

If the planes (1) and (2) are parallel, then two things can happen:

(a) *If* $(A_{11},A_{12},A_{13}) = a(A_{21},A_{22},A_{23})$ and *also* $b_1 = ab_2$, *then* equations (1) and (2) are exactly the same plane and any x in this plane satisfies both (1) and (2).

(b) *If* $(A_{11},A_{12},A_{13}) = a(A_{21},A_{22},A_{23})$ but $b_1 \neq ab_2$, *then* equations (1) and (2) are two distinct parallel planes and there is *no* point x which lies in both planes.

If we take *all three* equations together any of the following things may happen. If the planes (1) and (2) are *not* parallel, they intersect in a line, so any solution to all three equations must, at least, be on that line of intersection of (1) and (2). But x must also lie in the plane (3). Now the line of intersection of (1) and (2) *might* be parallel to the plane (3).

If it is *not*, then the line will pass through the plane (3) in eactly one point and that point x is then the *only* solution to the system of equations (1), (2), and (3); see Figure 4.1. If the line of intersection of (1) and (2) *is* parallel to (3) but *does not lie in the plane* (3), then there are *no* solutions to the system [(1),(2),(3)]; see Figure 4.2. If the line of intersection of (1) and (2) *lies in the plane* (3), then *any point x on that line* also satisfies (3) and is a solution of the system [(1),(2),(3)]; see Figure 4.3.

If the planes (1) and (2) *are* parallel, then in condition (a) (with $b_1 = ab_2$) we could have one of three things happen: the plane (3) is not parallel to (1) and (2) and so intersects them in a line and any point x on that line is a solution to the system [(1),(2),(3)]; see Figure 4.4; or the plane (3) is parallel to but not the same as (1) and (2), then there are no solutions to the

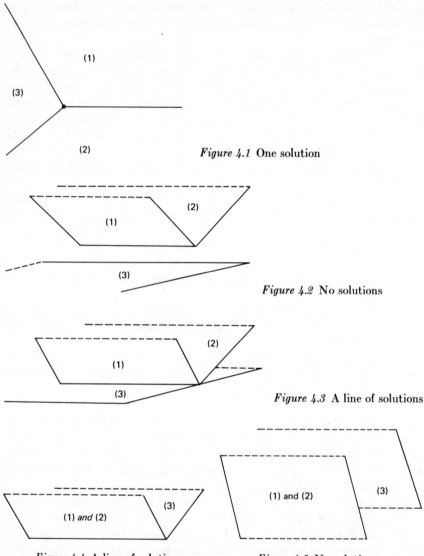

Figure 4.1 One solution

Figure 4.2 No solutions

Figure 4.3 A line of solutions

Figure 4.4 A line of solutions   *Figure 4.5* No solutions

system shown in Figure 4.5; or the plane (3) is the same as (1) and (2) and *any x* in that plane is a solution to the system (see Figure 4.6).

Finally, in condition (b) (with $b_1 \neq ab_2$) there are no solutions to the system, no matter where the plane (3) is (see Figure 4.7).

To apply the Gaussian elimination method to the system [(1),(2),(3)] we proceed as follows. We suppose that one of the coefficients A_{11}, A_{21}, and A_{31}

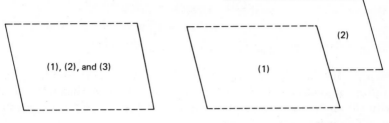

Figure 4.6 A line of solutions   Figure 4.7 No solutions

is not zero; otherwise there are no solutions or else a line or plane of solutions. Say $A_{11} \neq 0$; otherwise *reorder* the equations. (That is, if $A_{11} = 0$, but say $A_{31} \neq 0$, then exchange equations (1) and (3) and rename the coefficients so that with the new names, $A_{11} \neq 0$.)

Step 1 Subtract

$$\frac{A_{21}}{A_{11}} \text{ (1) from (2)}$$

and subtract

$$\frac{A_{31}}{A_{11}} \text{ (1) from (3).}$$

This replaces the system $[(1),(2),(3)]$ by the *equivalent* (with the same solutions) system.

(4) $\qquad\qquad A_{11}x_1 + A_{12}x_2 + A_{13}x_3 = b_1$

(5) $\qquad \left(A_{22} - \frac{A_{21}}{A_{11}} A_{12}\right)x_2 + \left(A_{23} - \frac{A_{21}}{A_{11}} A_{13}\right)x_3 = b_2 - \frac{A_{21}}{A_{11}} b_1$

(6) $\qquad \left(A_{32} - \frac{A_{31}}{A_{11}} A_{12}\right)x_2 + \left(A_{33} - \frac{A_{31}}{A_{11}} A_{13}\right)x_3 = b_3 - \frac{A_{31}}{A_{11}} b_1.$

We can write formulas for this transformation of equations (1), (2), and (3) into (4), (5), and (6) as follows.
    Define

$$A_{ij}^{(1)} = A_{ij}, \, b_i^{(1)} = b_i \qquad (i,j = 1,2,3)$$

and define

$$A_{ij}^{(2)} = A_{ij}^{(1)} - \frac{A_{i1}^{(1)}}{A_{11}^{(1)}} A_{1j}^{(1)} \qquad \text{for } i \geq 2; j \geq 2$$

$$b_i^{(2)} = b_i^{(1)} - \frac{A_{i1}^{(1)}}{A_{11}^{(1)}} b_1^{(1)} \qquad \text{for } i \geq 2.$$

Using this notation, (4), (5), and (6) become

$$\text{(4)} \qquad A_{11}^{(1)} x_1 + A_{12}^{(1)} x_2 + A_{13}^{(1)} x_3 = b_1^{(1)}$$
$$\text{(5)} \qquad A_{22}^{(2)} x_2 + A_{23}^{(2)} x_3 = b_2^{(2)}$$
$$\text{(6)} \qquad A_{32}^{(2)} x_2 + A_{33}^{(2)} x_3 = b_3^{(2)}$$

Step 2   If both $A_{22}^{(2)}$ and $A_{32}^{(2)}$ are zero, then there are no solutions or a line or plane of solutions. [If $A_{22}^{(2)}$ *is* zero, but $A_{32}^{(2)}$ is *not* zero, then interchange (5) with (6) and rename the coefficients so that the *new* $A_{22}^{(2)}$ is not zero.] Suppose $A_{22}^{(2)} \neq 0$.

Subtract $[A_{32}^{(2)}/A_{22}^{(2)}]$ (5) from (6) to obtain the equivalent system

$$\text{(7)} \qquad A_{11}^{(1)} x_1 + A_{12}^{(1)} x_2 + A_{13}^{(1)} x_3 = b_1^{(1)}$$
$$\text{(8)} \qquad A_{22}^{(2)} x_2 + A_{23}^{(2)} x_3 = b_2^{(2)}$$
$$\text{(9)} \qquad A_{33}^{(3)} x_3 = b_3^{(3)}$$

where

$$A_{ij}^{(3)} = A_{ij}^{(2)} - \frac{A_{i2}^{(2)}}{A_{22}^{(2)}} A_{2j}^{(2)} \qquad \text{for } i \geq 3, j \geq 3$$

$$b_i^{(3)} = b_i^{(2)} - \frac{A_{i2}^{(2)}}{A_{22}^{(2)}} b_2^{(2)} \qquad \text{for } i \geq 3.$$

The transformed system, [(7),(8),(9)], is now said to be in *upper triangular form*.

Step 3   We now perform the part of *Gaussian elimination* called *back substitution*. We solve equation (9) for x_3:

$$x_3 = \frac{b_3^{(3)}}{A_{33}^{(3)}}$$

(which we can do, provided $A_{33}^{(3)} \neq 0$; if $A_{33}^{(3)} = 0$, then there are no solutions if $b_3^{(3)} \neq 0$ or a line or plane of solutions if $b_3^{(3)} = 0$).

We substitute this value of x_3 into equations (7) and (8); we can then find x_2 from (8) as

$$x_2 = \frac{1}{A_{22}^{(2)}} [b_2^{(2)} - A_{23}^{(2)} x_3]$$

Finally, we substitute this value of x_2 into (7) and obtain x_1 as

$$x_1 = \frac{1}{A_{11}^{(1)}} [b_1^{(1)} - A_{12}^{(1)} x_2 - A_{13}^{(1)} x_3].$$

EXERCISES

1.   Find all the solutions (if any) of the systems:
   (a) $x_1 + x_2 + x_3 = 1$
   $2x_1 + 3x_2 + x_3 = 2$
   $3x_1 + 4x_2 + 2x_3 = 3$

(b) $x_1 + x_2 \qquad = 1$

$\qquad\quad 3x_2 + x_3 = 2$

$\quad 3x_1 \qquad + 2x_3 = 3$

(c) $x_1 + x_2 + x_3 = 0$

$\quad 2x_1 + 4x_2 + x_3 = 1$

$\quad x_1 + 3x_2 + x_3 = 1$

(d) $x_2 + x_3 + x_4 = 5$

$\quad x_1 + x_3 + x_4 = 4$

$\quad x_1 + x_2 + x_4 = 6$

$\quad x_1 + x_2 + x_3 = 6$

2. Write down the formulas for the Gaussian elimination method in the general case for $Ax = b$, with A a given $n \times n$ matrix, and b a given n-dimensional vector.

In case the problem arises to solve $Ax = b$ for a given $n \times n$ matrix A, but where b is going to run through a set of different n-tuples of numbers, then it might seem more efficient to find the inverse of A and compute the different solutions for the different b's by evaluating the formula

$$x = A^{-1}b$$

for the different b's. However, this is not the case, as is shown, for example, by the careful and thorough analysis of *operational* counts done by Isaacson and Keller, 1966. As they point out, "... since the final reduced matrix ... is upper triangular, we may store the multipliers $m_{i,k-1} = (A_{i,k-1}^{(k-1)}/A_{k-1,k-1}^{(k-1)})$, $2 \leq k \leq i \leq n$, in the lower triangular part of the original matrix, A (in the location of $A_{i,k-1}$)." In this way, to solve $Ax = b$ for a new vector b, we need not repeat the upper triangularization of A (it has already been done and the results are saved). We need form only the modified right-hand sides using the multipliers now stored in the lower triangular part of A. Thus

$$b_i^{(1)} = b_i \qquad (i = 1, 2, \ldots, n)$$

$$b_i^{(2)} = b_i^{(1)} - m_{i,1}b_1^{(1)} \qquad (i = 2, \ldots, n)$$

$$b_i^{(r+1)} = b_i^{(r)} - m_{i,r}b_r^{(r)} \qquad (i = r+1, \ldots, n) \qquad \text{for } r = 1, 2, \ldots, n-1.$$

The solution can then be completed for the new vector b by carrying out the back-substitution procedure. Counting only multiplications and divisions, Isaacson and Keller, 1966, find that it requires, in this way,

$$\frac{n^3}{3} + mn^2 - \frac{n}{3}$$

operations to solve $Ax = b$ for a fixed matrix A and a set of m different vectors b by Gaussian elimination.

But it requires n^3 operations to get A^{-1} (taking into account the zeros on the right-hand side; see Section 4.3) by Gaussian elimination and so

$n^3 + mn^2$ operations to solve the set of m equations $Ax = b$ by first finding A^{-1} and then multiplying A^{-1} into m different vectors to obtain m solutions of the form $x = A^{-1}b$.

4.3 Gaussian Elimination for A^{-1}

We can find A^{-1} by Gaussian elimination. To do this, we can use the Gaussian elimination method described in Section 4.2 to solve the *set* of equations

$$Ax^{(i)} = e_i \qquad (i = 1, 2, \ldots, n)$$

where e_i is the unit vector along the ith coordinate axis: the ith component of e_i is one and the other components are zero.

$$
e_i = \begin{pmatrix} 0 \\ 0 \\ \cdot \\ \cdot \\ \cdot \\ 0 \\ 1 \\ 0 \\ \cdot \\ \cdot \\ \cdot \\ 0 \end{pmatrix} \qquad i\text{th position.}
$$

The resulting solutions $x^{(i)}$ will be the *columns* of the inverse matrix, A^{-1}, thus

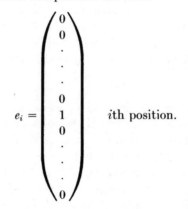

$$
A^{-1} = \begin{pmatrix} \cdot & & & & \cdot \\ \cdot & & & & \cdot \\ \cdot & & & & \cdot \\ x^{(1)} & \cdot & \cdot & \cdot & x^{(n)} \\ \cdot & & & & \cdot \\ \cdot & & & & \cdot \\ \cdot & & & & \cdot \end{pmatrix}.
$$

In other words,

$$(A^{-1})_{ij} = (x^{(j)})_i.$$

To illustrate, consider

$$A = \begin{pmatrix} 0 & 1 \\ -1 & 0 \end{pmatrix}.$$

To find A^{-1} by Gaussian elimination, we put

$$\begin{pmatrix} 0 & 1 \\ -1 & 0 \end{pmatrix} x^{(1)} = \begin{pmatrix} 1 \\ 0 \end{pmatrix}$$

and

$$\begin{pmatrix} 0 & 1 \\ -1 & 0 \end{pmatrix} x^{(2)} = \begin{pmatrix} 0 \\ 1 \end{pmatrix}.$$

In terms of components, we have, for the first of these systems

$$0 \cdot x_1^{(1)} + 1 \cdot x_2^{(1)} = 1$$
$$-1 \cdot x_1^{(1)} + 0 \cdot x_2^{(1)} = 0;$$

hence we put (exchanging the order of the equations)

$$-1 \cdot x_1^{(1)} \qquad = 0$$
$$1 \cdot x_2^{(1)} = 1$$

and we can solve for $x_2^{(1)}$ and $x_1^{(1)}$ getting

$$x^{(1)} = \begin{pmatrix} 0 \\ 1 \end{pmatrix}.$$

For the second system, we have

$$0 \cdot x_1^{(2)} + 1 \cdot x_2^{(2)} = 0$$
$$-1 \cdot x_1^{(2)} + 0 \cdot x_2^{(2)} = 1$$

and we can obtain

$$x^{(2)} = \begin{pmatrix} -1 \\ 0 \end{pmatrix}.$$

Thus we have

$$A^{-1} = \begin{pmatrix} \cdot & \cdot \\ x^{(1)} & x^{(2)} \\ \cdot & \cdot \\ \cdot & \cdot \end{pmatrix} = \begin{pmatrix} 0 & -1 \\ 1 & 0 \end{pmatrix}.$$

To check this, we can verify that

$$AA^{-1} = \begin{pmatrix} 0 & 1 \\ -1 & 0 \end{pmatrix}\begin{pmatrix} 0 & -1 \\ 1 & 0 \end{pmatrix} = \begin{pmatrix} 1 & 0 \\ 0 & 1 \end{pmatrix} = I.$$

EXERCISES

1. Find the inverse of the matrix

$$A = \begin{pmatrix} 1 & 1 & 2 \\ 1 & 2 & 3 \\ 2 & 3 & 1 \end{pmatrix}$$

Use the inverse to find the solutions of

$$Ax = b$$

for

(a) $b = \begin{pmatrix} 8 \\ 11 \\ 15 \end{pmatrix}$

(b) $b = \begin{pmatrix} 10 \\ 1 \\ -3 \end{pmatrix}$

(c) $b = \begin{pmatrix} 0 \\ 1 \\ 0 \end{pmatrix}$

(d) $b = \begin{pmatrix} 1 \\ 1 \\ 1 \end{pmatrix}$

2.   Count the number of operations (omitting multiplications by zero) required by Gaussian elimination method to find A^{-1} for a 3×3 matrix A (saving the upper triangularization and multipliers).

4.4   Iteration for $Ax = b$

We can put

$$f(x) = Ax - b$$

and seek a vector x such that $f(x) = 0$.

Proceeding as in Section 2.1 on *iterative methods*, we can set the function to be iterated as

$$g(y) = y + h(y)f(y)$$

and choose h so that $g'(y)$ is near zero when y is near a zero of f. We have, in this case,

$$g(y) = y + h(y)(Ay - b)$$

so (see Section 3.1)

$$g'(y) = I + h'(y)(Ay - b) + h(y)A.$$

To get $g'(y)$ near zero when y is near a solution to $Ax = b$, we want $h(y)$ to be near $-A^{-1}$. In fact, if we knew A^{-1}, then putting $h(y) = -A^{-1}$, we would get

$$g(y) = y - A^{-1}(Ay - b) = y - Iy + A^{-1}b$$
$$= A^{-1}b.$$

Using this g, with *any* initial guess $y^{(0)}$, we get, using the iteration

$$y^{(k+1)} = g(y^{(k)})$$

the *exact* result

$$y^{(1)} = g(y^{(0)}) = A^{-1}b$$

in one iteration.

But it is harder to get A^{-1} than to solve $Ax = b$ in the first place, so this is an impractical method.

Suppose the $n \times n$ matrix A has no zero elements on its *main diagonal*, $A_{ii} \neq 0$ for $i = 1, 2, \ldots, n$. Let D_A be the diagonal matrix whose *diagonal elements* $(D_A)_{ii}$ are the same as those of A, namely A_{ii}, and whose other elements are zero.

Thus

$$(D_A)_{ij} = \begin{cases} A_{ii} & \text{for } i = j \\ 0 & \text{for } i \neq j. \end{cases}$$

Clearly, D_A^{-1} can be written down at once:

$$(D_A^{-1})_{ij} = \begin{cases} \dfrac{1}{A_{ii}} & \text{for } i = j \\ 0 & \text{for } i \neq j. \end{cases}$$

The system $Ax = b$ is equivalent to the system

$$(D_A^{-1}A)x = D_A^{-1}b.$$

Furthermore, $D_A^{-1}A$ can be written as

$$D_A^{-1}A = I + M$$

where M has only zeros on its diagonal (so D_M is a matrix of all zeros). For example, if

$$A = \begin{pmatrix} 2 & 1 \\ 1 & 4 \end{pmatrix},$$

then

$$D_A = \begin{pmatrix} 2 & 0 \\ 0 & 4 \end{pmatrix}$$

$$D_A^{-1} = \begin{pmatrix} \frac{1}{2} & 0 \\ 0 & \frac{1}{4} \end{pmatrix}$$

$$D_A^{-1}A = \begin{pmatrix} 1 & \frac{1}{2} \\ \frac{1}{4} & 1 \end{pmatrix}$$

$$= \begin{pmatrix} 1 & 0 \\ 0 & 1 \end{pmatrix} + \begin{pmatrix} 0 & \frac{1}{2} \\ \frac{1}{4} & 0 \end{pmatrix}.$$

Now if, when we do all this, it turns out that the matrix M has all of its elements small compared to 1, then a reasonable approximation to $(D_A^{-1}A)^{-1}$ can be obtained as

$$(D_A^{-1}A)^{-1} = (I + M)^{-1} \approx I - M.$$

To check this in the example above, notice that

$$I - M = \begin{pmatrix} 1 & -\frac{1}{2} \\ -\frac{1}{4} & 1 \end{pmatrix}$$

whereas

$$(D_A^{-1}A)^{-1} = \begin{pmatrix} 1 & \frac{1}{2} \\ \frac{1}{4} & 1 \end{pmatrix}^{-1} = \begin{pmatrix} \frac{8}{7} & -\frac{4}{7} \\ -\frac{2}{7} & \frac{8}{7} \end{pmatrix}.$$

Thus, using $h(y) = -(I - M)$, we can try the iterative method

$$y^{(k+1)} = g(y^{(k)})$$

with

$$g(y^{(k)}) = y^{(k)} - (I - M)(D_A^{-1}Ay^{(k)} - D_A^{-1}b)$$

where

$$M = D_A^{-1}A - I.$$

Thus g can be written as

$$g(y^{(k)}) = [I - (I - M)D_A^{-1}A]y^{(k)} + (I - M)D_A^{-1}b.$$

EXERCISES

1. Show that $g(y^{(k)})$ can also be written in the simpler form $g(y^{(k)}) = M^2 y^{(k)} + (I - M)D_A^{-1}b$.

2. Show that the same sequence $(y^{(k)})$ can be generated by two steps at a time of the still simpler iteration formula

$$y^{(k+1)} = -My^{(k)} + D_A^{-1}b.$$

3. Show that the vector function $f(y) = -My + D_A^{-1}b$ is a contraction mapping on all of E^n if $\|M\| < 1$ for any matrix norm. Conclude that the iterative method will converge from *any starting vector* $y^{(0)}$ if $\|M\| < 1$; for instance, for the *maximum row sum norm*

$$\|M\| = \max_{i=1, 2, \ldots, n} \sum_{J=1}^{n} |M_{ij}|.$$

Let us try the method on $Ax = b$ with

$$A = \begin{pmatrix} 2 & 1 \\ 1 & 4 \end{pmatrix} \quad \text{and} \quad b = \begin{pmatrix} 1 \\ 2 \end{pmatrix}.$$

We have, as before,

$$D_A^{-1}A = \begin{pmatrix} 1 & \frac{1}{2} \\ \frac{1}{4} & 1 \end{pmatrix}$$

and we find that

$$D_A^{-1}b = \begin{pmatrix} \frac{1}{2} & 0 \\ 0 & \frac{1}{4} \end{pmatrix}\begin{pmatrix} 1 \\ 2 \end{pmatrix} = \begin{pmatrix} \frac{1}{2} \\ \frac{1}{2} \end{pmatrix}$$

and

$$(I - M)D_A^{-1}b = \begin{pmatrix} 1 & -\frac{1}{2} \\ -\frac{1}{4} & 1 \end{pmatrix}\begin{pmatrix} \frac{1}{2} \\ \frac{1}{2} \end{pmatrix} = \begin{pmatrix} \frac{1}{4} \\ \frac{3}{8} \end{pmatrix}$$

and

$$[I - (I - M)D_A^{-1}A] = \begin{pmatrix} 1 & 0 \\ 0 & 1 \end{pmatrix} - \begin{pmatrix} 1 & -\frac{1}{2} \\ \frac{1}{4} & 1 \end{pmatrix}\begin{pmatrix} 1 & \frac{1}{2} \\ \frac{1}{4} & 1 \end{pmatrix} = \begin{pmatrix} \frac{1}{8} & 0 \\ 0 & \frac{1}{8} \end{pmatrix}$$

and so the iteration formula becomes, in this example,

$$g(y^{(k)}) = \begin{pmatrix} \frac{1}{8} & 0 \\ 0 & \frac{1}{8} \end{pmatrix}y^{(k)} + \begin{pmatrix} \frac{1}{4} \\ \frac{3}{8} \end{pmatrix}.$$

If we take

$$y^{(0)} = \begin{pmatrix} 0 \\ 0 \end{pmatrix}$$

as a starting guess, and compute $y^{(1)}, y^{(2)}, \ldots$ from $y^{(k+1)} = g(y^{(k)})$, we obtain

$$y^{(1)} = \begin{pmatrix} \frac{1}{8} & 0 \\ 0 & \frac{1}{8} \end{pmatrix}\begin{pmatrix} 0 \\ 0 \end{pmatrix} + \begin{pmatrix} \frac{1}{4} \\ \frac{3}{8} \end{pmatrix} = \begin{pmatrix} \frac{1}{4} \\ \frac{3}{8} \end{pmatrix}$$

$$y^{(2)} = \begin{pmatrix} \frac{1}{8} & 0 \\ 0 & \frac{1}{8} \end{pmatrix}\begin{pmatrix} \frac{1}{4} \\ \frac{3}{8} \end{pmatrix} + \begin{pmatrix} \frac{1}{4} \\ \frac{3}{8} \end{pmatrix} = \begin{pmatrix} 0.28 \ldots \\ 0.39 \ldots \end{pmatrix}$$

$$y^{(3)} = \begin{pmatrix} \frac{1}{8} & 0 \\ 0 & \frac{1}{8} \end{pmatrix}\begin{pmatrix} 0.28 \\ 0.39 \end{pmatrix} + \begin{pmatrix} \frac{1}{4} \\ \frac{3}{8} \end{pmatrix} = \begin{pmatrix} 0.285 \ldots \\ 0.423 \ldots \end{pmatrix}$$

$$y^{(4)} = \begin{pmatrix} 0.2856 \ldots \\ 0.4278 \ldots \end{pmatrix}.$$

The exact solution is

$$x = \begin{pmatrix} \frac{2}{7} \\ \frac{3}{7} \end{pmatrix} = \begin{pmatrix} .2857 \ldots \\ .4285 \ldots \end{pmatrix}.$$

4.5   Iteration for A^{-1}

To find a matrix X such that $f(X) = AX - I = 0$, we can again (in case the diagonal elements of A are not zero) put

$$D_A^{-1}A = I + M$$

and put

$$\begin{aligned} g(Y) &= Y - (I - M)(D_A^{-1}AY - D_A^{-1}) \\ &= Y - (I - M)((I + M)Y - D_A^{-1}) \\ &= M^2Y + (I - M)D_A^{-1}. \end{aligned}$$

The iteration to find A^{-1} based on this form of g is then

$$Y^{(k+1)} = M^2Y^{(k)} + (I - M)D_A^{-1}.$$

Let us try this for

$$A = \begin{pmatrix} 2 & 1 \\ 1 & 4 \end{pmatrix}.$$

We have

$$D_A^{-1} = \begin{pmatrix} \frac{1}{2} & 0 \\ 0 & \frac{1}{4} \end{pmatrix}$$

as before and

$$M^2 = \begin{pmatrix} 0 & \frac{1}{2} \\ \frac{1}{4} & 0 \end{pmatrix}^2 = \begin{pmatrix} \frac{1}{8} & 0 \\ 0 & \frac{1}{8} \end{pmatrix}$$

so

$$Y^{(k+1)} = \begin{pmatrix} \frac{1}{8} & 0 \\ 0 & \frac{1}{8} \end{pmatrix}Y^{(k)} + \begin{pmatrix} \frac{1}{2} & -\frac{1}{8} \\ -\frac{1}{8} & \frac{1}{4} \end{pmatrix}.$$

EXERCISE

Put

$$Y^{(0)} = \begin{pmatrix} 0 & 0 \\ 0 & 0 \end{pmatrix}$$

and find $Y^{(k+1)}$ from

$$Y^{(k+1)} = \begin{pmatrix} \frac{1}{8} & 0 \\ 0 & \frac{1}{8} \end{pmatrix}Y^{(k)} + \begin{pmatrix} \frac{1}{2} & -\frac{1}{8} \\ -\frac{1}{8} & \frac{1}{4} \end{pmatrix} \qquad \text{for } k = 0, 1, 2.$$

How may iterations (what value of k) are needed before $Y^{(k+1)}$ is within .001 of A^{-1} element by element?

4.6 Accuracy, Efficiency, and Storage Requirements

4.6.1 Accuracy

With *exact* arithmetic and *exact* initial data, Gaussian elimination would give *exact* solutions to linear algebraic systems. (Iterative methods would not.) With round-off error present as in limited precision machine arithmetic, and with inexact initial data, the question arises: How accurate are the numerical solutions of linear algebraic equations obtained by Gaussian elimination?

In Section 1.6 we discussed the effects of rounding errors on the accuracy of numerical solutions. The effect is, that for any given *nonsingular* square matrix A, the system $Ax = b$ (for a given vector b) can be solved for x using the Gaussian elimination method; if N binary place machine arithmetic is used (with a basic relative rounding error of 2^{-N}, low-order bit), then the relative error (in the norm $\|x\|_\infty$) of the numerical solution for x will be bounded by $C \cdot 2^{-N}$, where C is some number independent of N.

This means that for a given nonsingular matrix A and vector b there is a large enough N (double precision, triple precision, or whatever) so that we can have a numerical solution by Gaussian elimination as accurate as we please. If we want a relative error in

$$\|x\|_\infty = \max_{i=1, 2, \ldots, n} |x_i|$$

less than ϵ we need only take N large enough so that

$$C \cdot 2^{-N} < \epsilon.$$

However, the factor C does depend on the matrix A, the vector b, and on certain details of the algorithm used. For instance, an interchange of two of the rows of A, corresponding to an interchange in the order of two equations in the system $Ax = b$ (say we interchange the first and second equations), may change the value of C. Similarly interchanging two columns of A, corresponding to renaming the components of x (say we interchange the order of x_2 and x_3), may change the value of C. These row and column interchanges are referred to as *pivoting*. If pivoting is used during Gaussian elimination, the value of C will be reduced as much as possible when the *complete pivoting strategy* is used, selecting the largest remaining coefficient as the next diagonal element (after interchanges) during the upper triangularization process (Forsythe and Moler, 1967). Even then C is going to depend on the *size* of the matrix A and on its *condition*.

To discuss the notion of the *condition* of a matrix, we first recall from Section 1.6 that the computed solution $\bar{x}$ (Gaussian elimination with pivoting) satisfies $(A + \delta A)\bar{x} = b$ exactly for some *error matrix* δA with $\|\delta A\|_\infty$ *rarely larger than* $n2^{-N}\|A\|_\infty$ $(u = 2^{-N}$, being the unit round off). On the other hand, this does *not* mean that we can put $C = n\|A\|_\infty$. In fact, if

$$(A + \delta A)\bar{x} = b$$

and

$$Ax = b,$$

then the error vector

$$\delta x = x - \bar{x}$$

satisfies

$$A\delta x = (\delta A)\bar{x}$$
$$\delta x = (A^{-1}\delta A)\bar{x}$$

so that we can only say that

$$\frac{\|\delta x\|_\infty}{\|\bar{x}\|_\infty} \leq (\|A^{-1}\|_\infty\|A\|_\infty) \frac{\|\delta A\|_\infty}{\|A\|_\infty}.$$

Thus, although $\|\delta A\|_\infty/\|A\|_\infty$ might be only n units of round off, we could have $\|\delta x\|_\infty/\|\bar{x}\|_\infty$ larger than this by the factor cond $(A) = \|A^{-1}\|_\infty\|A\|_\infty$, called the *condition number* of A. This is sometimes expressed in terms of the Euclidean norm; then cond $(A) = \|A^{-1}\| \; \|A\| = \mu_1/\mu_n \geq 1$, where μ_1 and μ_n are, respectively, the moduli of the largest and smallest eigenvalues of A. If A is singular, or *nearly singular*, then cond (A) is infinite or very large and we say that A is *ill-conditioned*. Notice this is not the same thing as a small value for the determinant of A. In fact, det A is the product of eigenvalues and so the diagonal matrix

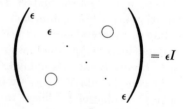

has determinant ϵ^n but condition number 1.

EXERCISE

Show that the *nearly singular* system

$$x_1 + x_2 = 1$$
$$x_1 + (1 + \epsilon)x_2 = b_2 \quad (\epsilon \text{ small})$$

is ill-conditioned; that is, the matrix

$$\begin{pmatrix} 1 & 1 \\ 1 & 1 + \epsilon \end{pmatrix}$$

is ill-conditioned when ϵ is small compared to 1. The two equations in the system represent *nearly parallel* lines (they become parallel when $\epsilon = 0$). If ϵ is very small, then when b_2 is changed by just a little, the point of intersection of the two lines will change by a great deal. Show this with a diagram.

The condition of a matrix can sometimes be improved by *scaling* or *equilibration* (Forsythe and Moler, 1967); for example, if ϵ is a small positive number and

$$A = \begin{pmatrix} \epsilon & 0 \\ 0 & \dfrac{1}{\epsilon} \end{pmatrix},$$

we can rescale the linear system

$$\epsilon x_1 \qquad\quad = b_1$$

$$\frac{1}{\epsilon} x_2 = b_2$$

by multiplying the second equation by ϵ^2, then we have the scaled system

$$\epsilon x_1 \qquad = b_1$$

$$\epsilon x_2 = \epsilon^2 b_2$$

with coefficient matrix

$$A' = \begin{pmatrix} \epsilon & 0 \\ 0 & \epsilon \end{pmatrix}.$$

Now cond $(A) = (1/\epsilon)^2$, whereas cond $(A') = 1$. Note that $\det(A) = 1$, whereas det $(A') = \epsilon^2$.

Iterative improvement (Forsythe and Moler, 1967) of a solution $\bar{x}$ obtained by Gaussian elimination (or by any other procedure) can be made if *residuals* are calculated with enough precision. We put $x^{(0)} = \bar{x}$, and calculate (*with high precision*) the *residual vector*

$$r_m = b - A x^{(m)}$$

and then solve for $\delta x^{(m)}$ in the equation

$$A \delta x^{(m)} = r_m$$

and then compute the *improved* solution

$$x^{(m+1)} = x^{(m)} + \delta x^{(m)}.$$

Iterative improvement can be combined with interval arithmetic to yield realistic, strict bounds on round-off error (Hansen, 1965; Hansen and Smith, 1967; Hansen, 1969; Moore, 1966). We carry out the final iteration only in interval arithmetic with high precision.

The situation is more complicated with respect to interval versions of *direct* methods such as Gaussian elimination. Von Neumann and Goldstine (1947, p. 1023) state that "Matrices of order 15, 50, 150 can usually be inverted with a (relative) precision of 8, 10, 12 decimal digits less respectively than the number of digits carried throughout." These are the kinds of bounds that have been obtained using rounded interval arithmetic with Gaussian eliminations, as the numerical experience that we will cite will show. The statement quoted is based on an *a priori* analysis of rounding error for Gaussian elimination and so shares some aspects of *worst-case analysis* (Wilkinson, 1963) with interval computation.

On the other hand, actual computing experience with test cases of large or moderate-sized matrices since 1947 has shown that often one obtains much better results than are predicted by such "worst-case analyses."

For interval calculations, "the midpoint of an interval produced by interval arithmetic is often much more accurate than the width of the interval indicates" (Richman, Dec. 3, 1969). This "midpoint phenomenon" does not always occur, but does often enough that it would be very useful to find some way of taking advantage of it to reduce the computed error bounds.

Numerical experience with interval versions of Gaussian elimination (and variants) includes the following.

A study was carried out by Collins, 1960 on the IBM 704 computer using interval arithmetic to invert some 35 matrices of orders from 5 to 20 whose elements were chosen at random from the interval $[-1,1]$. Collins reported that: "For these matrices, the average number of significant decimal digits in an element of the ith row of the inverse turned out to be approximately $8.13 - .37n - .15(n - i)$, n being the order of the matrix." Note that for $n = 15$ and $i = 1$ this *conclusion* predicts a loss of about 7.65 decimal digits — in close agreement with von Neumann's figure.

In another study (Moore, 1962) it was reported that with interval Gaussian elimination (using about 8 decimals) on the IBM 7090 computer, "some 14×14 matrices were inverted with resulting intervals containing the coefficients of the inverse matrix of relative width about 10^{-1}." This loss of about 7 decimals for $n = 14$ is again in good agreement with von Neumann's figures. It is also shown (Moore, 1962, p. 102) that *in any rounded interval arithmetic computation beginning with exact initial data and carrying N binary digits, the width of the resulting intervals will be bounded by some constant independent of N times* 2^{-N}. In other words, the number of digits lost should be essentially independent of the number carried and depend only on the particular matrix. Thus, for high enough machine precision even Gaussian elimination round-off errors for a given matrix can be strictly bounded by interval computation with intervals as narrow as one pleases. Double-precision interval arithmetic (say 16 to 20 decimals) should give narrow bounds for most matrices of moderate order.

On the other hand, when the coefficients of a matrix to be inverted (or a system of linear equations to be solved) are known only approximately, then we have to consider the propagation of initial error as well as the accumulation of round-off error. A fundamental difficulty arises here because of the fact that the set of possible exact solutions is geometrically complicated and is not simply a K-tuple of intervals. Hansen discusses this in a recent paper (1969, pp. 35–46).

EXERCISES

1. Show that (Forsythe and Moler, 1967) if

$$Ax = b$$

and

$$A(x + \delta x) = b + \delta b,$$

then

$$\frac{\|\delta x\|}{\|x\|} \leq \text{cond } (A) \, \frac{\|\delta b\|}{\|b\|}.$$

2. Show that (Forsythe and Moler, 1967)

$$\frac{\|B^{-1} - A^{-1}\|}{\|B^{-1}\|} \leq \text{cond } (A) \, \frac{\|A - B\|}{\|A\|}.$$

4.6.2   Efficiency

There are two basically different types of methods for solving linear algebraic equations: (1) *direct* methods, such as Gaussian elimination, which except for rounding error do give exact results in a finite number of arithmetic operations and (2) *indirect* or *iterative* methods which do not.

To solve one nth-order linear algebraic system

$$Ax = b$$

using Gaussian elimination requires about $n^3/3$ operations, in fact $n^3/3 + n^2 - n/3$ (Issacson and Keller, 1966).

If the system can be just as easily written in the form

$$(I + M)x = b$$

with $\|M\| < 1$, then we can consider using an iterative method, for instance,

$$y^{(k+1)} = -My^{(k)} + b.$$

If $\|M\| = c < 1$, then (see Section 2.2 on convergence of iterative methods) putting $y^{(0)} = 0$, we have

$$\|x - y^{(k)}\| \leq \frac{c^k}{1 - c}\|b\|.$$

Thus the relative error in $y^{(k)}$ as an approximation to x is about c^k. To make $c^k \leq 10^{-5}$, for instance, would require

$$k \geq \frac{5 \ln 10}{\ln (1/c)}.$$

EXERCISE

Show that

$$\frac{\|b\|}{1 + c} \leq \|x\| \leq \frac{\|b\|}{1 - c}.$$

Now each iteration is going to take n^2 operations if M is *dense* (very few zero elements) but only sn operations if M is *sparse* (mostly zero elements) with about s nonzero elements in each row. (For instance, for a tridiagonal matrix, $s = 3$.) Thus, to achieve at least five-decimal-digit accuracy in each element of $y^{(k)}$, we would need to perform about

$$\frac{11.5}{\ln \left(\dfrac{1}{\|M\|}\right)} \quad sn \text{ operations.}$$

Using machine arithmetic with a precision of around 18 decimals (double-precision is often around this), we should be able to achieve this accuracy with Gaussian elimination for most matrices of order to at least 150 (probably 200 or more). Let us say that *for* $n = 200$ we can achieve at least five-decimal-place accuracy (for most matrices) in the solution of $Ax = b$, using Gaussian elimination with pivoting by using a machine arithmetic of precision equivalent to about N binary places ($N \approx 56$?) without taking any advantage of the presence of any possible zero elements in the matrix. Gaussian elimination will require about $n^3/3 \approx 2.7 \cdot 10^6$ operations, whereas iteration will require about

$$\frac{11.5 \, n^2}{\ln \left(\dfrac{1}{\|M\|}\right)} \approx \frac{4.6 \cdot 10^5}{\ln \left(\dfrac{1}{\|M\|}\right)}$$

so that iteration will require probably fewer operations (when $n = 200$, if $\|M\| < e^{-.17} \approx .84$) than Gaussian elimination by about the ratio $.17/[\ln (1/\|M\|)]$; if $\|M\|$ is small, this can be substantial.

If we do take advantage of the presence of many zeros in the matrix A (for instance, as happens in finite difference methods for solving boundary-value problems (Forsythe and Moler, 1960; Varga, 1962; Young, 1971), we may have perhaps $s = 3$ or $s = 5$), then iteration gains an extra advantage since we need only multiply nonzero elements of M into $y^{(k)}$. It is not so easy (or always possible) to make a similar savings in Gaussian elimination by taking advantage of sparseness.

However, there is one special case when a substantial savings over the usual $n^3/3$ operations can be made, namely, when the matrix is tridiagonal. In this case the number of operations in Gaussian elimination can be trimmed to $5n - 4$ (Issacson and Keller, 1966, p.57).

The convergence of iteration methods can be accelerated by various means, such as, for instance, "over relaxation" (Forsythe and Moler, 1967; Issacson and Keller, 1966; Varga, 1962; Young, 1971).

EXERCISES

1. Compare the number of operations required by *Cramer's rule* with that required by Gaussian elimination.
2. Devise an efficient way to evaluate determinants.
3. Reconsider Exercise 1.
4. How does Gaussian elimination simplify for a tridiagonal system ($A_{ij} = 0$ for $|i - j| > 1$)?

4.6.3   Storage Requirements

The first thing that must be *stored* in a computer to solve a linear system of algebraic equations is, of course, a *program* for the solution. There are many good programs in existence. Forsythe and Moler, 1967, give programs written in the FORTRAN, ALGOL 60, Extended ALGOL, and PL/1 languages. More recently a volume of computer programs has appeared in the Springer handbook series (Wilkinson and Reinsch, 1971). Most computing centers will be equipped with a variety of "canned routines" for solving linear algebraic systems. To paraphrase a common philosophy: *Let the user beware* of the possibility of *ill-conditioned* systems and of the need for sufficient precision in machine arithmetic — especially for *large* systems.

Very often the coefficients in a linear algebraic system to be solved on a computer need not be stored at all as an array but can be *generated* easily as they are needed during a computation. This is especially true for iterative methods when, often, we need store only the approximate solution vector. In fact, a variant on the iteration method discussed so far in this chapter, which is generally called *simultaneous displacements* or the *Jacobi iteration method*, is that of *successive displacements* or the *Gauss-Seidel iteration method*.

With the Jacobi method we must save the entire vector $y^{(k)}$ until we have completed the computation of

$$y^{(k+1)} = -My^{(k)} + b.$$

In component form, this means that, from $y_j^{(k)}$, $j = 1, 2, \ldots, n$ we compute and save

$$y_i^{(k+1)} = -\sum_{j=1}^{n} M_{ij}y_j^{(k)} + b_i \qquad \text{for } i = 1, 2, \ldots, n.$$

Thus we need storage for both the vectors $y^{(k)}$ and $y^{(k+1)}$.

The Gauss-Seidel method enables us to cut this storage requirement in half. We simply use the *current* value of each component as we obtain it. That is, as soon as we compute $y_i^{(k+1)}$, we store it "on top of" $y_i^{(k)}$. Thus we *successively* displace the components of the *old* iterate $y^{(k)}$ by those of the *new* iterate $y^{(k+1)}$. Formally, we then have

$$y_i^{(k+1)} = -\sum_{j=1}^{i-1} M_{ij}y_j^{(k+1)} - \sum_{j=i+1}^{n} M_{ij}y_j^{(k)} + b_i$$

(since $M_{ij} = 0$, the term $j = i$ does not appear). In this way the vectors $y^{(k)}$ and $y^{(k+1)}$ share the same storage space in the Gauss-Seidel method. The analysis of convergence for this iterative method is similar to but somewhat more complicated than for the Jacobi method (Forsythe and Wasow, 1960; Forsythe and Moler, 1967; Issacson and Keller, 1966; Varga, 1962; Young, 1971). The methods are comparable with respect to accuracy in most cases.

4.7 Linear Programming

Most practical problems that require computing are not simply presented in the form of a single equation of some standard type to be solved numerically. Rather, there are usually various inequalities, constraints, side conditions, and what not to be satisfied; and some collection of formulas and equations describing (approximately) relations between measurable quantities, along with some measured values of some of the variables, constants, parameters, or whatever. Usually, it is possible to sort all this out to some extent at least so that a number of *subproblems* can be identified and treated. By putting together a collection of algorithms for computing solutions to the subproblems in the proper way, we can build up a *system* for handling a complicated computing job.

Inequalities may be used to define regions, for instance, restricting an argument of a function to part of the whole domain of the function. For example, the inequality

$$|z| < 1$$

is satisfied by points z in the complex plane lying inside the unit circle. And the inequality

$$f(x) \geq 3$$

can be used to define a subset of the domain of f, namely, those argument values for which the function value is greater than or equal to 3. We might have a problem in which we are looking for an x that satisfies a number of equations and restrictions, one of which is that x is in the set

$$\{x \mid f(x) \geq 3\}.$$

If x is supposed to satisfy simultaneously a number of such conditions, each confining x to a certain region, then the combined effect of the collection of conditions is to confine x to the *intersection* of the separate regions. For example, the pair of inequalities

$$x < 3$$

and

$$x > 1$$

taken together restricts x to the intersection of the two regions: (1) the numbers to the left of 3 on the real line; (2) the numbers to the right of 1 on the real line. The intersection of regions (1) and (2) is the set of numbers between 1 and 3: $\{x \mid 1 < x < 3\}$. Thus the conjunction of inequalities corresponds to the intersection of regions. Similarly, the disjunction of inequalities (one inequality *or* another must be satisfied) corresponds to the *union* of regions (x is in one region *or* another.)

A *linear programming problem* is one in which we seek to *maximize* (or *minimize*) a certain linear combination, $c_1x_1 + c_2x_2 + \cdots + c_nx_n$, of numerical values of quantities $x_1, x_2, \ldots, x_n$ subject to a finite set of *linear inequalities* (constraints)

$$a_{1,1}x_1 + a_{1,2}x_2 + \cdots + a_{1,n}x_n \leq b_1$$
$$a_{2,1}x_1 + a_{2,2}x_2 + \cdots + a_{2,n}x_n \leq b_2$$
$$\vdots$$
$$a_{k,1}x_1 + a_{k,2}x_2 + \cdots + a_{k,n}x_n \leq b_n.$$

The number k of constraints does not have to be the same as n. It can be much greater, for example. Each of the separate linear relations defines a region of points in E^n and the collection of inequalities taken together defines a region that is the intersection of the separate regions. The problem is to find a point in this *admissible* region that maximizes the given linear combination (Dantzig, 1963).

To illustrate, suppose we wish to find a point (x_1,x_2) in the plane which maximizes the linear function $p(x_1,x_2) = (\frac{1}{3})x_1 + (\frac{2}{3})x_2$ subject to the constraints

$$-x_1 + x_2 \leq 1$$
$$3x_1 + 15x_2 \leq 10$$
$$2x_1 + x_2 \leq 2.$$

Each relation confines the point (x_1,x_2) to the region (in this case) *below* the line corresponding to the case of equality in that relation (the bounding line is given by $-x_1 + x_2 = 1$ for the first region). The intersection of the three regions is bounded by an open polygon (see Figure 4.8). Only the points in the heavily shaded region satisfy all three constraints.

The *level curves* of the function to be maximized are straight lines of the form

$$p(x_1,x_2) = \text{constant};$$

that is,

$$(\tfrac{1}{3})x_1 + (\tfrac{2}{3})x_2 = c.$$

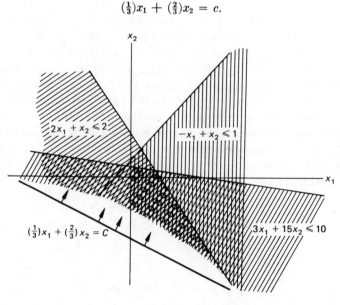

Figure 4.8

One such curve is shown as the heavy line. As c increases the line moves up parallel to the one shown. We want to choose the biggest c for which the heavy line still has a point in the heavily shaded region (of points which satisfy all the constraints). From Figure 4.8 it can be seen that this will occur when c is such that the heavy line

$$(\tfrac{1}{3})x_1 + (\tfrac{2}{3})x_2 = c$$

goes through the point of intersection of the lines

$$2x_1 + x_2 = 2$$

and

$$3x_1 + 15x_2 = 10.$$

The solution to the problem is the point whose coordinates satisfy this system of linear algebraic equations, namely

$$(x_1, x_2) = \left(\frac{20}{27}, \frac{14}{27}\right).$$

EXERCISE

The following exercise is taken from E. Stiefel, *An Introduction to Numerical Mathematics*. New York: Academic Press, 1963, p. 22.

A farmer owns 100 acres of land. He wants to plant potatoes in one part, corn in another, and perhaps leave the rest fallow. We also have the following information:

	Potatoes	Corn	Total Available
Cultivation costs in dollars per acre	10	20	1100
Workdays per acre	1	4	160
Net profit in dollars per acre	40	120	

The figures in the last column indicate that the farmer has a capital of $1100 and 160 workdays to spend. Now, how is he to organize his planting so as to realize a maximum net profit?

If x acres are planted with potatoes and y acres with corn, the first two rows of the table state that

$$10x + 20y \leq 1100, \qquad x + 4y \leq 160.$$

Also $\qquad x + y \leq 100 \qquad$ and $\quad x \geq 0, y \geq 0.$

The problem of maximizing the net profit can be formulated as follows: find (x,y) satisfying all five constraints and such that

$$40x + 120y$$

is maximized.

Just as the conjunction of inequalities corresponds to the intersection of regions, so the conjunction of equations corresponds to the intersection of solutions, regarding solutions as sets of points.

Each of the equations

$$-x_1 + x_2 = 1$$
$$3x_1 + 15x_2 = 10$$
$$2x_1 + x_2 = 2$$

taken separately has a line of points in the (x_1,x_2) plane as a solution. The first two of them taken together (the *conjunction*) have as a joint (or simultaneous) solution the point at which the corresponding two lines intersect, namely the point $(-5/18,13/18)$. The second and third equations taken together intersect in the point $(20/27,14/27)$. We could ask for the points that satisfy *either* the first two equations *or* the second and third equations and the answer would be that there are *two* such points (the ones we have just described). If we ask for the points which satisfy all three equations together, the answer is that there are *none*. The three lines do not come together at any point.

The equation

$$x_1^2 + x_2^2 + x_3^2 = 1$$

describes a spherical surface of points in E^3. An equation of the form

$$ax_1 + bx_2 + cx_3 = 0$$

where a, b, and c are given real numbers describes a plane through the origin in E^3. Such a plane will intersect the spherical surface, just given, in a circle whose orientation in E^3 depends on a, b, and c.

EXERCISE

Find all points (x_1,x_2) which satisfy the system

$$x_1^2 + x_2^2 = 1$$
$$x_1^2 - x_2^2 = 0$$
$$x_2 > 0.$$

(*Hint:* Graph the curves and find numerical values for the coordinates (x_1,x_2) of intersection points.)

apter 5

Continuous approximation

5.1 Interpolation

If f is a *discrete* function with domain $T = (x_1, x_2, \ldots, x_n)$ consisting of a finite set of real numbers, then a continuous function g with domain (a,b) containing T is called an *extension* of f if $g(x_i) = f(x_i)$ for $i = 1, 2, \ldots, n$.

Clearly there are many extensions of a given discrete function. In fact, if g is an extension of f and f has domain T, then $g + h$ is also an extension of f for any function h that vanishes on T. Put another way, any function g whose graph passes through the points $[x_1, f(x_1)], \ldots, [x_n, f(x_n)]$ is an extension of f (see Figure 5.1). A continuous extension of a discrete function is

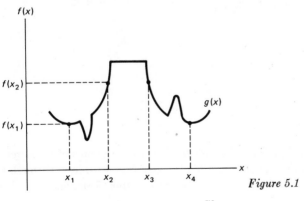

Figure 5.1

said to *interpolate* (pass through) the values (points) of the discrete function.

Piecewise linear interpolation is useful for estimating values of differentiable functions from tables. For instance, in a table of values of the exponential function e^x we find the successive entries:

x	e^x
. . .	. . .
. . .	. . .
2.02	7.5383
2.03	7.6141
2.04	7.6906
. . .	. . .
. . .	. . .

Figure 5.2

Of course, e^x is a continuous (and differentiable) function defined for all real x. We could use a variety of methods for computing e^x for any x directly particularly on a computer without using this or any other table. On the other hand, such tables are still useful for giving a quick approximation without having to go to the computer for only a number or two.

To get a value of e^x for some x not in the table by piecewise linear interpolation, we continuously connect successive points in the table by line segments. For x between two successive entries $x_i < x < x_{i+1}$ we approximate e^x by

$$e^x \approx e^{x_i} + \frac{e^{x_{i+1}} - e^{x_i}}{x_{i+1} - x_i} (x - x_i).$$

For instance, if $x = 2.0231$, we have

$$e^{2.0231} \approx 7.5383 + \frac{7.6141 - 7.5383}{2.03 - 2.02} (2.0231 - 2.02)$$

or

$$e^{2.0231} \approx 7.5618.$$

EXERCISE

Use Figure 5.2 to approximate $e^{2.039}$.

Just as there is a unique line through *two* successive values of a discrete function, so there is a unique polynomial of degree $k - 1$ (or less) through k successive values. It is called the *Lagrange interpolation formula*.

If f is a discrete function with domain T and if $x_1, x_2, \ldots, x_k$ are *distinct* elements of T, then the system of equations

$$p(x_1) = a_0 + x_1 a_1 + \cdots + x_1^{k-1} a_{k-1} = f(x_1)$$
$$p(x_2) = a_0 + x_2 a_1 + \cdots + x_2^{k-1} a_{k-1} = f(x_2)$$

.
.
.

$$p(x_k) = a_0 + x_k a_1 + \cdots + x_k^{k-1} a_{k-1} = f(x_k)$$

has a unique solution for the coefficients $a_0, a_1, \ldots, a_{k-1}$ of a polynomial $p_{k-1}(x) = a_0 + a_1 x + \cdots + a_{k-1} x^{k-1}$ which interpolates the values of f at $x_1, x_2, \ldots, x_k$.

EXERCISE

Find the polynomial $p_2(x) = a_0 + a_1 x + a_2 x^2$ which interpolates the three values of e^x given in Figure 5.2.

There is often an interesting *tradeoff* between efficiency and storage requirements in the design of a *good* computer subroutine for evaluating commonly used functions — such as exponential. We could store a large number of values and use a simple and quick interpolation for intermediate values. On the other hand, we could store few or no values but use a more lengthy, elaborate algorithm for generating them.

EXERCISE

Design a *good* subroutine for computing e^x for x in [0,1]. Give a quantitative discussion of the average (or maximum) number of operations required versus the storage required. What would seem a reasonable compromise?

There is a vast mathematical literature on interpolation and *approximation theory* that cannot be even sketched here; but the most important mathematical topic in approximation theory for scientific computing is certainly *least squares*. We will go into this in some depth in this chapter.

5.2 Splines

The term *spline* dates back to devices and techniques used in the design of ships and in other construction problems in which it is desired to find a very *smooth* curve passing through several points. In recent years, I. J. Schoenberg and others have developed a very elegant mathematical theory of spline approximation (Ahlberg et al., 1967; Greville, 1969; Kimeldorf and Wahba, 1971; Schoenberg, 1967).

Suppose we wish to find a very smooth curve passing through the points shown in Figure 5.3. We could try a succession of *French curves* and perhaps find one that fitted well; or we could sketch a curve "by hand" ("by eye"), but if we want a *formula* as well that will enable us to calculate the coordinates of points on a smooth interpolating curve, then we need a more mathematical approach.

Suppose we put the points in an x-y coordinate system as shown in Figure 5.3 and determine by measurement that the coordinates of the four given points are $(-1,0)$, $(0,0)$, $(1,0.1)$, $(1.3,1)$ or, in tabular form,

x	y
-1	0
0	0
1	0.1
1.3	1

Now we *could* find the Lagrange interpolation formula for the four points that would give a third degree (cubic) polynomial passing through the four points. Let us do this now and plot the resulting curve to see what it looks like.

Referring to Section 5.1, we write (using the table of coordinate values for the four points)

$$a_0 + (-1)a_1 + (-1)^2a_2 + (-1)^3a_3 = 0$$
$$a_0 + \quad (0)a_1 + \quad (0)^2a_2 + \quad (0)^3a_3 = 0$$
$$a_0 + \quad (1)a_1 + \quad (1)^2a_2 + \quad (1)^3a_3 = 0.1$$
$$a_0 + (1.3)a_1 + (1.3)^2a_2 + (1.3)^3a_3 = 1.$$

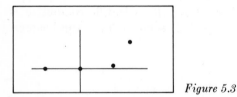

Figure 5.3

We can solve for a_0, a_1, a_2, a_3 to obtain the interpolating polynomial $p(x) = a_0 + a_1x + a_2x^2 + a_3x^3$. We find that $a_0 = 0$, $a_1 = -.898$, $a_2 = 0.05$, $a_3 = .948$, and the Lagrange interpolating polynomial passes through the four given points, as shown in Figure 5.4. Now, the interpolating polynomial shown *is* a *smooth* curve in the sense that it has no jumps or sharp corners; however, it has a large *oscillation* between $x = -1$ and $x = +1$. Suppose we want a still smoother interpolating function — one without the large oscillation *and* without jumps or sharp corners (a *broken-line* or *piecewise linear* interpolation will have sharp corners).

One approach is to use *spline functions*, which are piecewise polynomials with matching slopes for adjoining pieces and even matching higher derivatives at the points of interpolation.

For the example under discussion (Figure 5.3) we can seek a spline function s of degree 2 satisfying the following conditions:

(1)
$$s(-1) = 0$$
$$s(0) \; = 0$$
$$s(1) \; = 0.1$$
$$s(1.3) = 1$$

(2)
$$s(x) = \begin{cases} p_1(x) = a_0^{(1)} + a_1^{(1)}x + a_2^{(1)}x^2 & \text{for } -1 \le x \le 0 \\ p_2(x) = a_0^{(2)} + a_1^{(2)}x + a_2^{(2)}x^2 & \text{for } \;\; 0 \le x \le 1 \\ p_3(x) = a_0^{(3)} + a_1^{(3)}x + a_2^{(3)}x^2 & \text{for } \;\; 1 \le x \le 1.3. \end{cases}$$

Thus we require that $p_1(0) = p_2(0)$ and $p_2(1) = p_3(1)$ or

$$a_0^{(1)} = a_0^{(2)}$$
$$a_0^{(2)} + a_1^{(2)} + a_2^{(2)} = a_0^{(3)} + a_1^{(3)} + a_2^{(3)};$$

and

(3)
$$\frac{dp_1(0)}{dx} = \frac{dp_2(0)}{dx} \quad \text{and} \quad \frac{dp_2(1)}{dx} = \frac{dp_3(1)}{dx}$$

or

$$a_1^{(1)} = a_1^{(2)}$$
$$a_1^{(2)} + 2a_2^{(2)} = a_1^{(3)} + 2a_2^{(3)}.$$

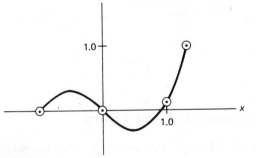

$y = -.898x + 0.05x^2 + .948x^3$

Figure 5.4

Putting together conditions (1), (2), and (3), we obtain the following equations for $a_0^{(1)}$, $a_1^{(1)}$, $a_2^{(1)}$, $a_0^{(2)}$, $a_1^{(2)}$, $a_2^{(2)}$, $a_0^{(3)}$, $a_1^{(3)}$, $a_2^{(3)}$:

$$a_0^{(1)} - a_1^{(1)} + a_2^{(1)} = 0$$
$$a_0^{(1)} = 0$$
$$a_0^{(2)} = 0$$
$$a_0^{(2)} + a_1^{(2)} + a_2^{(2)} = 0.1$$
$$a_0^{(3)} + a_1^{(3)} + a_2^{(3)} = 0.1$$
$$a_0^{(3)} + 1.3a_1^{(3)} + (1.3)^2 a_2^{(3)} = 1$$
$$a_1^{(1)} - a_1^{(2)} = 0$$
$$a_1^{(2)} + 2a_2^{(2)} - a_1^{(3)} - 2a_2^{(3)} = 0.$$

There are eight equations in nine unknowns and we may choose one of the unknowns arbitrarily. We can put $a_2^{(1)} = 0$, for instance, to get a straight-line segment $p_1(x) \equiv 0$ for $-1 \le x \le 0$ interpolating the first two data points. Then we can solve for a unique set of values for the remaining co-efficients to obtain the spline function

$$s(x) = \begin{cases} p_1(x) \equiv 0 & \text{for } -1 \le x \le 0 \\ p_2(x) = 0.1x^2 & \text{for } 0 \le x \le 1 \\ p_3(x) = 9.2 - 18.4x + 9.3x^2 & \text{for } 1 \le x \le 1.3. \end{cases}$$

We can verify that this s satisfies conditions (1), (2), and (3), and Figure 5.5 shows the graph of $s(x)$ interpolating the data points with no large oscillations.

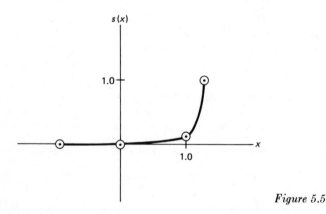

Figure 5.5

5.3   Discrete Linear-Least-Squares Curve Fitting

The simple but eminently useful technique discussed in this section may enjoy the widest application of any technique in mathematics outside arithmetic, elementary geometry, and trigonometry. In fact, it is related to elementary geometry as will be shown in subsequent sections.

A function whose domain is finite is called a *discrete* function. We can view an n-dimensional vector $x = (x_1, x_2, \ldots, x_n)$ as the discrete function

$$\{(1, x_1), (2, x_2), \ldots, (n, x_n)\}$$

with domain $\{1, 2, \ldots, n\}$ and with function values $x(1) = x_1, \ldots,$ $x(n) = x_n$.

Conversely, for a given domain, we can view a discrete function as an n-dimensional vector, where n is the number of elements in the domain of the function. Consider, for instance, a discrete function x *given in tabular form* (a "table" of values) as

t	$x(t)$
t_1	$x(t_1)$
t_2	$x(t_2)$
t_3	$x(t_3)$
$\ldots$	$\ldots$
t_n	$x(t_n)$

or *given in graphical form* as a set of points in the plane (see Figure 5.6). For a fixed choice of the domain $T = \{t_1, t_2, \ldots, t_n\}$, we can represent discrete functions by n-dimensional vectors; for instance,

$$x = [x(t_1), x(t_2), \ldots, x(t_n)].$$

Given another discrete function y with the same domain T, we can use the vector representation to define a *distance* (which we could *compute*) between the two functions x and y as

$$\|x - y\| = \left\{ \sum_{k=1}^{n} (x(t_k) - y(t_k))^2 \right\}^{1/2}.$$

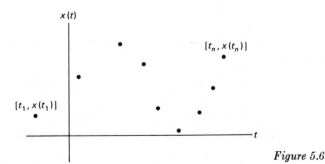

Figure 5.6

EXERCISES

1. Compute the distance between the discrete functions x and y given by the following table of values:

t	$x(t)$	$y(t)$
0.1	1.1	1.3
0.2	0.9	1.0
0.8	9.7	9.5

2. Write down a function whose distance from x is half that of y.
3. Let x and z be discrete functions with common domain $\{0.1, 0.2, 0.8,\}$ where x is given as in Exercise 1. Suppose that $\|x - z\| \leq 0.1$. What is the interval of possible values for $z(0.1)$?

Using this definition of distance, consider now the following problem.

Let the discrete function x be given in tabular form with domain T. Let φ_1 and φ_2 be two functions whose domains both include all the distinct points $t_1, t_2, \ldots, t_n$ in T. Consider the *two-parameter family* of functions of the form $y = a\varphi_1 + b\varphi_2$, defined by

$$y(t) = a\varphi_1(t) + b\varphi_2(t)$$

whose domains are the intersection of the domains of φ_1 and φ_2.

We now define an operation on functions whose domains include T. We call the operation $|_T$ defined by

$$f|_T = \{(t, f(t)) | t \in T\}$$

the *restriction* of f to T. This is an example of *discretization* (about which we will have more to say later).

Thus

$$\varphi_1|_T = \{[t_1, \varphi_1(t_1)], \ldots, [t_n, \varphi_1(t_n)]\} \quad \text{and}$$

$$\varphi_2|_T = \{[t_1, \varphi_2(t_1)], \ldots, [t_n, \varphi_2(t_n)]\}$$

are discrete functions with domain T.

We have $\varphi_1|_T(t_i) = \varphi_1(t_i)$ $(i = 1, 2, \ldots, n)$ and $\varphi_2|_T(t_i) = \varphi_2(t_i)$ $(i = 1, 2, \ldots, n)$.

The operation $f|_T$ simply *restricts* the domain of f to the set T and throws away the rest of the function f.

Clearly $(a\varphi_1 + b\varphi_2)|_T = a\varphi_1|_T + b\varphi_2|_T$. We will now consider the following problems: for what values of the parameters a and b does the discrete function $y|_T$ have the smallest distance from the given function x?

To illustrate, consider the discrete function x as given in the previous exercise. Let φ_1 and φ_2 be, respectively, the constant function $\varphi_1(t) \equiv 1$ and the linear function given by $\varphi_2(t) = t$ for all t. The question then becomes: for what values of parameters a and b does the discrete function $y|_T$, obtained by restricting $y(t) = a\varphi_1(t) + b\varphi_2(t) = a + bt$ to the finite set $T = \{0.1, 0.2, 0.8\}$, have the smallest distance from x? We can compute that

$$\|x - y|_T\| = \left(\sum_{k=1}^{3} (x(t_k) - y|_T(t_k))^2 \right)^{1/2}$$

$$= \{(1.1 - a - b(0.1))^2 + (0.9 - a - b(0.2))^2$$
$$+ (9.7 - a - b(0.8))^2\}^{1/2}.$$

The solution to this problem gives the *best fit* (in the sense of *least squares*) to the data (the table of values of x) by any function y *of the form chosen*.

The parameter values for the best fit can be found as follows. We want to minimize (among all choices of a and b)

$$\|x - y\| = \|x - (a\varphi_1 + b\varphi_2)|_T\|$$

by an appropriate *best* choice of a and b.

We have

$$\|x - (a\varphi_1 + b\varphi_2)|_T\| = \left(\sum_{k=1}^{n} (x(t_k) - (a\varphi_1(t_k) + b\varphi_2(t_k)))^2 \right)^{1/2}.$$

This can be put into the form

$$D^2(a,b) = c_0 - 2c_1 a - 2c_2 b + A_{11}a^2 + 2A_{12}ab + A_{22}b^2$$

where

$$D^2(a,b) = \|x - (a\varphi_1 + b\varphi_2)|_T\|^2$$

$$c_0 = \sum_{k=1}^{n} (x(t_k))^2$$

$$c_1 = \sum_{k=1}^{n} x(t_k)\varphi_1(t_k)$$

$$c_2 = \sum_{k=1}^{n} x(t_k)\varphi_2(t_k)$$

$$A_{11} = \sum_{k=1}^{n} (\varphi_1(t_k))^2$$

$$A_{12} = \sum_{k=1}^{n} \varphi_1(t_k)\varphi_2(t_k)$$

and

$$A_{22} = \sum_{k=1}^{n} (\varphi_2(t_k))^2.$$

For the example under discussion these expressions become

$$c_0 = \sum_{k=1}^{3} (x(t_k))^2 = (1.1)^2 + (0.9)^2 + (9.7)^2 = 96.11$$

$$c_1 = \sum_{k=1}^{3} x(t_k)\varphi_1(t_k) = (1.1)(1) + (0.9)(1) + (9.7)(1) = 11.7$$

$$c_2 = \sum_{k=1}^{3} x(t_k)\varphi_2(t_k) = (1.1)(0.1) + (0.9)(0.2) + (9.7)(0.8) = 8.05$$

$$A_{11} = \sum_{k=1}^{3} [\varphi_1(t_k)]^2 = 1^2 + 1^2 + 1^2 = 3$$

$$A_{12} = (1)(0.1) + (1)(0.2) + (1)(0.8) = 1.1$$

$$A_{22} = (0.1)^2 + (0.2)^2 + (0.8)^2 = .69$$

If $(\varphi_1(t_1), \ldots, \varphi_1(t_n))$ is not a scalar multiple of $(\varphi_2(t_1), \ldots, \varphi_2(t_n))$, then $A_{11}A_{22} > A_{12}{}^2$ and it follows that $D^2(a,b)$ is then a nonnegative quadratic function of a and b, and it will have a minimum value. Since the square root function is monotonic (if $s > t$, then $\sqrt{s} > \sqrt{t}$) for nonnegative arguments, it follows that $D(a,b)$ will also have its minimum value at the same a, b. We can picture $D^2(a,b)$ as a surface over the (a,b) plane, as in Figure 5.7.

At the minimum value of $D^2(a,b)$ the values of a and b will satisfy the equations

$$\frac{\partial D^2(a,b)}{\partial a} = 0 \quad \text{and} \quad \frac{\partial D^2(a,b)}{\partial b} = 0.$$

To find a and b, then, we solve the equations

$$-2c_1 + 2A_{11}a + 2A_{12}b = 0$$
$$-2c_2 + 2A_{12}a + 2A_{22}b = 0$$

(which we can do if $A_{11}A_{12} > A_{12}{}^2$).

For our illustrative example, $A_{11} = 3$, $A_{22} = .69$, and $A_{12} = 1.1$; and we can check that $A_{11}A_{22} > A_{12}{}^2$. The system of equations to solve for the best values of a and b then becomes (omitting the superfluous 2's)

$$-11.7 + 3a + 1.1b = 0$$
$$-8.05 + 1.1a + .69b = 0.$$

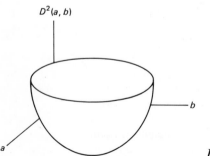

$D^2(a, b)$

b

a

Figure 5.7

EXERCISES

1. Solve the above system of linear algebraic equations for a and b. Plot the resulting best-fitting straight line $y(t) = a + bt$ to the data represented by the discrete function $x = \{(0.1,1.1), (0.2,0.9), (0.8,9.7)\}$. Plot the data points also; and compute the distance of this y from x ($y|_r$, that is). Compare this distance with the distance obtained for the y in the previous set of Exercises. Explain why this distance is not less than that of the y in the previous Exercises.

2. Find the straight lines that best fit (in the least-squares sense) the following sets of data. Include graphs showing *both* the data and the line found.

(a)

t	$x(t)$
-7	3
-5	2
0	1
3	0
8	-1

(b)

t	$x(t)$
1	3
2	2
3	3
4	6

3. Write a computer program for finding the coefficients a, b of the least squares line for an *arbitrary* set of data points $\{(t_1,x(t_1)), (t_2,x(t_2)), \ldots, (t_n,x(t_n))\}$.

4. Find the best least-squares, straight-line fit $\{\varphi_1(t) = 1, \varphi_2(t) = t\}$ to the data given by the discrete function $x = \{(1,2), (2,1), (2.5,1.5), (4,0.5)\}$. Plot the result.

Remarks on Applications

The range of possible *applications* of least-squares linear approximation is enormous and it would be difficult to give a complete account of it. (The same is true for many of the other mathematical and computing techniques described in this book, but perhaps not to the same extent.)

It might be interesting to give here at least some faint inkling of what is meant by these last remarks by listing a set of applications of the least-squares, straight-line fit to a set of data points in the plane. The students in one undergraduate course (at the University of Wisconsin were asked to apply the least-squares technique *to some data of interest to them* and the following variety of applications resulted:

1. Cloud height versus vertical wind speed
2. Tornado diameter versus number of hours passed
3. Number of revolutions of a wheel produced by an initial acceleration versus number of minutes passed
4. Velocity versus time (of a falling object)
5. Absorption versus reddening (for Galactic Cluster III Cepheus and for Galactic Cluster h and χ Persei)
6. Sales versus advertising costs (for XYZ corporation, 1966–1971)
7. Sales versus Government estimate of industry demand (for "ABC" corporation, 1950–1959)
8. Amount of O_2 evolved through photosynthesis by a sprig of Elodea submerged in 2 percent aqueous $NaHCO_3$ at three different distances from a light source of fixed intensity
9. Annual consumer expenditures versus annual net income
10. Processing time for a product versus quality of product
11. Comparison of age distribution in an underdeveloped country with that in a developed country
12. Mixed cost versus machine hours (for two different machines)
13. Property value versus square feet
14. Property value versus distance from central shopping district
15. pH-reading versus cerium added
16. Absorbance versus EDTA added
17. Personal consumption expenditure versus gross national product
18. Personal savings versus gross national product
19. Earnings per share of common stock versus calendar year
20. Minimum temperature versus previous day's minimum temperature
21. Mathematics test scores versus verbal test scores
22. High temperature versus low temperature
23. Logarithm of concentration of $(CH_3)_3CBr$ versus time
24. Mean of liquid and vapor densities versus temperature (for methyl ether and CCl_4)
25. Wage earnings in building construction versus calendar year
26. Passenger miles on domestic airlines versus calendar year
27. Swimming speed versus percentage time with extended pectoral fins
28. Number of people living in London during 1693 versus years of age
29. Mass difference between mirror nuclei versus $(Z_1 - 1)A^{-1/3}$ as determined from an equation using positron decay energy between two isobars

5.4 Discrete Nonlinear Least Squares

In fitting a curve to data points there are two main things to consider. First, what *kind* of curve do we want? Second, having somehow decided what *kind* of curve we want, how do we get the numerical values of coefficients or parameters appearing in curves of that *kind* so that we have the *best fitting* curve of that kind to the data points? We must define *best fitting*, of course.

Whereas in Section 5.3 we considered, as fitting curves, linear combinations of two functions, in this section we will consider fitting curves of any form in which there occur a finite number of parameters $p_1, p_2, \ldots, p_M$ to be determined.

Suppose that data $\{(x_1, f(x_1)), \ldots, (x_n, f(x_n))\}$ are given representing a discrete function f with domain $T = \{x_1, x_2, \ldots, x_n\}$. Let $\varphi_p(x) = \varphi(x, p_1, p_2, \ldots, p_M)$ be an *M-parameter family* of curves of some chosen form whose domains include all x in T, where $p = (p_1, p_2, \ldots, p_M)$ is an M-tuple of parameter values,

$$ f \equiv (f(x_1), f(x_2), \ldots, f(x_n)) $$

and

$$ \varphi_{p|T} \equiv (\varphi_p(x_1), \varphi_p(x_2), \ldots, \varphi_p(x_n)). $$

We wish to find $\bar{p}$ such that $\|f - \varphi_{\bar{p}|T}\| \le \|f - \varphi_{p|T}\|$ for all p.

We can *visualize* the problem geometrically in Euclidean n-space E^n in the following way:

We are given a *point f*, the n-dimensional vector of data values $(f(x_1), f(x_2), \ldots, f(x_n))$; we seek a vector $\bar{p}$ of parameters for which the *point*

$$ \varphi_{\bar{p}|T} = (\varphi_{\bar{p}}(x_1), \varphi_{\bar{p}}(x_2), \ldots, \varphi_{\bar{p}}(x_n)) $$

is as close as possible to the point f among *all* points $\varphi_{p|T}$ we can get with any M-tuple of parameters $p = (p_1, p_2, \ldots, p_M)$. To do this, it is convenient to consider

$$ D(p) = \|f - \varphi_{p|T}\| = \left\{ \sum_{k=1}^{n} (f(x_k) - \varphi_p(x_k))^2 \right\}^{1/2} $$

as defining a *distance* function D whose domain is the *set* of different p's. For each $p = (p_1, p_2, \ldots, p_M)$ we have a corresponding point $\varphi_{p|T}$ in E^n, so the set of all such points $\{\varphi_{p|T} \mid p \in E^M\}$ is an M-dimensional *surface* Φ in E^n. If f lies in Φ, then $\varphi_{\bar{p}|T} = f$ for some $\bar{p}$ and $D(\bar{p}) = 0$. If f does not lie in Φ, then there may be one or more places on the surface Φ which are closest to f.

Suppose that $\varphi_p(x) = \varphi(x, p_1, p_2, \ldots, p_M)$ is differentiable (locally linear) in all its variables. In that case $D(\bar{p}) = D(p_1, p_2, \ldots, p_M)$ is also differentiable with respect to $p_1, p_2, \ldots, p_M$ and if $D(\bar{p}) \le D(p)$, we must have

$$\frac{\partial D(\bar{p})}{\partial p_1} = \frac{\partial D(\bar{p})}{\partial p_2} = \cdots = \frac{\partial D(\bar{p})}{\partial p_M} = 0.$$

This leads to the system of equations:

$$\sum_{k=1}^{n} [f(x_k) - \varphi(x_k,\bar{p}_1, \ldots, \bar{p}_M)] \frac{\partial \varphi(x_k,\bar{p}_1, \ldots, \bar{p}_M)}{\partial p_1} = 0$$

.

.

.

$$\sum_{k=1}^{n} [f(x_k) - \varphi(x_k,\bar{p}_1, \ldots, \bar{p}_M)] \frac{\partial \varphi(x_k,\bar{p}_1, \ldots, \bar{p}_M)}{\partial p_M} = 0.$$

This is a system of equations of the form $F(\bar{p}) = 0$. If $\varphi(x,p_1, \ldots, p_M)$ is a *linear* function of $p_1, \ldots, p_M$, then the system of equations will be linear in $\bar{p}_1, \ldots, \bar{p}_M$, and we can use Gaussian elimination to solve it.

EXERCISE

(Linear least squares) Let $\varphi_1(x)$, $\varphi_2(x)$, $\ldots$, $\varphi_M(x)$ be chosen and consider linear combinations

$$\varphi(x,p_1,p_2, \ldots, p_M) = p_1\varphi_1(x) + p_2\varphi_2(x) + \cdots + p_M\varphi_M(x)$$

with coefficients p_1, p_2, $\ldots$, p_M. Show that the best (least squares) values for the parameters p_1, p_2, $\ldots$, p_M are found by solving the linear system

$$\sum_{j=1}^{M} (\varphi_i,\varphi_j)\bar{p}_j = (\varphi_i,f) \qquad (i = 1,2, \ldots, M)$$

where the coefficients in the system are the *inner products:*

$$(\varphi_i,\varphi_j) = \sum_{k=1}^{n} \varphi_i(x_k)\varphi_j(x_k)$$

$$(\varphi_i,f) = \sum_{k=1}^{n} f(x_k)\varphi_i(x_k).$$

In this form, we can see that the difference between f and the linear least-squares approximation $\bar{p}_1\varphi_1 + \cdots + \bar{p}_M\varphi_M$ is *perpendicular* to (makes a zero inner product with) each of the vectors

$$\varphi_{i|_T} = (\varphi_i(x_1),\varphi_i(x_2), \ldots, \varphi_i(x_n)) \qquad (i = 1,2, \ldots, M).$$

If $f(x_k) - [\bar{p}_1\varphi_1(x_k) + \cdots + \bar{p}_M\varphi_M(x_k)] = g(x_k)$, then

$$(\varphi_i,g) = (\varphi_i,f) - \sum_{j=1}^{M} (\varphi_i,\varphi_j)\bar{p}_j = 0 \qquad (i = 1,2, \ldots, M).$$

If φ is nonlinear in $p_1, p_2, \ldots, p_M$, then $F(\bar{p}) = 0$ is a system of nonlinear equations with

$$F(\bar{p}) = F(\bar{p}_1, \bar{p}_2, \ldots, \bar{p}_M) = \begin{pmatrix} F_1(\bar{p}_1, \bar{p}_2, \ldots, \bar{p}_M) \\ F_2(\bar{p}_1, \bar{p}_2, \ldots, \bar{p}_M) \\ \cdot \\ \cdot \\ \cdot \\ F_M(\bar{p}_1, \bar{p}_2, \ldots, \bar{p}_M) \end{pmatrix}$$

where

$$F_i(\bar{p}_1, \ldots, \bar{p}_M) = \sum_{k=1}^{n} (f(x_k) - \varphi(x_k, \bar{p}_1, \ldots, \bar{p}_M)) \frac{\partial \varphi(x_k, \bar{p}_1, \ldots, \bar{p}_M)}{\partial p_i}.$$

We may use, for instance, Newton's method to solve the system if $F(p)$ is differentiable in p with nonsingular Jacobian matrix near the solution $\bar{p}$, and if we have a sufficiently good initial approximation $p^{(0)}$ to $\bar{p}$ so that the linear approximation to F near $p^{(0)}$ is *close to* the linear approximation to F near $\bar{p}$.

We are not, by any means, limited to Newton's method. A number of alternative numerical methods for finding a $\bar{p}$ that minimizes $D(p)$ can be considered. Some of them involve computing $D(p)$ for several different guesses, say $p^{(1)}, p^{(2)}, \ldots, p^{(q)}$, and then estimating the *direction* of "steepest descent" of $D(p)$ and using this information to get closer to $\bar{p}$ (Daniel, 1971).

EXERCISES

1. Write down the system of equations to be solved for the best (least squares) approximation of the form $\varphi(x, p_1, p_2) = e^{-p_1(x-p_2)^2}$ for a given set of data points $\{(x_1, f(x_1)), (x_2, f(x_2)), \ldots, (x_n, f(x_n))\}$.

2. Suppose an experiment yields a continuous recording in the form of a complicated looking oscillatory curve $D(t)$. Suppose further that we seek an approximate representation $A(t)$ [of the data $D(t)$] in the form

$$A(t) = \sum_{i=1}^{N} a_i \sin (w_i t + v_i).$$

(a) Describe, in detail, numerical procedures for finding (approximately) a best (in some sense) set of parameter values $a_1, a_2, \ldots, a_N$, $w_1, w_2, \ldots, w_N, v_1, v_2, \ldots, v_N$.

(b) What simplifications are possible if all the v_i's are *chosen* to be zero and all the w_i's are *chosen* to be multiples of $(2\pi/T)$?

If we are free to choose the *form* of the fitting curve, we can use the distance function, D, to choose between various possible forms. For instance, suppose we want to choose between fitting curves of two possible forms

$\varphi^{(1)}(x,p_1,p_2, \ldots, p_M)$ or $\varphi^{(2)}(x,q_1,q_2, \ldots, q_N)$. We can allow different numbers of parameters. We *could* base a choice between $\varphi^{(1)}$ or $\varphi^{(2)}$ upon a comparison of the values of $D(\bar{p})$ and $D(\bar{q})$, choosing whichever form of fitting curve gives the smallest value of the distance function. This requires first *solving* for the best parameter values $\bar{p}$ and $\bar{q}$ for each of the forms $\varphi^{(1)}$ and $\varphi^{(2)}$.

We turn now to a specific application of nonlinear least-squares curve fitting.

Recall from the discussion of an earth-Mars spaceflight in Chapter 1 that we had estimated we could reach Mars in a spaceship on an elliptical orbit around the sun by leaving the vicinity of earth, racing ahead of the earth in *its* orbit around the sun, with an escape velocity of around 6930 mph.

Now there arises the following practical problem. Assume we start our spaceship on a journey, hopefully to Mars, leaving the earth's vicinity with the velocity 6930 mph just as planned. Sometime later we will want to determine where we are and where we seem to be headed, and we will want to compute any necessary corrective thrusts we must make in order to reach Mars. This is in the nature of an interplanetary navigation problem.

There are, to be sure, many different kinds of instruments we could use to obtain information that would be helpful in determining our position in space. And there are many ways of computing our present course and of estimating corrections needed to reach our destination.

If we can make observations that will determine with some accuracy our position and velocity in the x_1, x_2 coordinate system described in Appendix A, then we could use the differential equations of motion for the spaceship

$$\ddot{x}_1 = \frac{-g}{r^2} \frac{x_1}{r}$$

$$\ddot{x}_2 = \frac{-g}{r^2} \frac{x_2}{r}$$

to predict our course by carrying out a numerical solution using a method such as described in Chapter 6. By repeating such a calculation several times for different initial velocities incorporating proposed corrective thrusts, we could determine a corrective thrust that would put us on a truer course toward a rendezvous with Mars. Another approach would be to make a number of observations to determine *positional* data along the flight and then find the elliptical orbit that best fits the data.

To get a fix on the position of the spaceship, we could measure a pair of angles. We can choose a distant identifiable star for this purpose and assume that the line of sight to the star from any position of the spacecraft during the flight will make a fixed angle, θ_0, with the positive x_1-axis of our (x_1,x_2) coordinate system (whose origin is at the sun).

We will suppose that at any time t the position (X_1,X_2) of the earth is known.

Suppose we *measure* the angle θ_1 between the line of sight (from the spaceship) to the star and the line of sight to the sun. Then the position of the spaceship (x_1,x_2) lies on the line through the sun which makes an angle of $\theta_0 + \theta_1$ with the positive x_1-axis. Suppose we also measure the angle θ_2 between the line of sight (from the spaceship) to the earth and the line of sight to the star. The position of the spaceship at that time must then lie on the line through the earth, which makes an angle θ_2 with any line to the star. We have found, then, two lines on which the position of the spaceship must lie and the point of intersection of the two lines is a fix on the position of the spacecraft (see Figure 5.8).

It is convenient to use *polar coordinates* r, θ in this calculation, for then we have, from Figure 5.8,

$$x_1 = r \cos \theta$$
$$x_2 = r \sin \theta$$

where

$$\theta = \theta_0 + \theta_1 - \pi$$

and

$$r = \frac{X_1 \sin (\theta_2 - \theta_0) + X_2 \cos (\theta_2 - \theta_0)}{\sin (2\pi - \theta_1 - \theta_2)}.$$

We can write the equation of an ellipse with a focus at the sun as

$$r(\theta,p_1,p_2,p_3) = \frac{p_1}{1 - p_2 \cos (\theta - p_3)}.$$

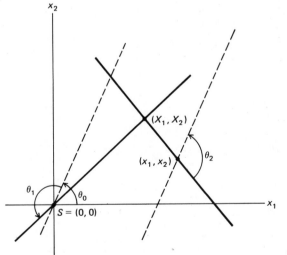

Figure 5.8

For instance, if $p_3 = \pi$ and $p_1 = 2ME/(E + M)$ and $p_2 = (M + E)/(E + M)$, then

$$r\left(0,\pi,\frac{2ME}{E + M},\frac{M - E}{E + M}\right) = E$$

and

$$r\left(\pi,\pi,\frac{2ME}{E + M},\frac{M - E}{E + M}\right) = M$$

and we have the *nominal* ellipse, namely the planned orbit of the spacecraft.

Suppose we have made a number of measurements of angles at various times during the first part of the flight and have fixed a set of positions during the course of the flight as $\{(\theta_1,r_1), (\theta_2,r_2), \ldots, (\theta_n,r_n)\}$ in θ, r coordinates. We seek the best-fitting ellipse in the form

$$r_p(\theta) = r(\theta,p_1,p_2,p_3) = \frac{p_1}{1 - p_2 \cos (\theta - p_3)}$$

to the observed positional data. For a given $p = (p_1,p_2,p_3)$, the distance function for the least-squares, curve-fitting problem here is

$$D(p) = \left\{\sum_{k=1}^{n} [r_k - r_p(\theta_k)]^2\right\}^{1/2}$$

and the system of equations to be solved for the best parameter values $\bar{p} = (\bar{p}_1,\bar{p}_2,\bar{p}_3)$ is

$$\sum_{k=1}^{n} [r_k - r_{\bar{p}}(\theta_k)] \frac{\partial r_{\bar{p}}(\theta_k)}{\partial p_1} = 0$$

$$\sum_{k=1}^{n} [r_k - r_{\bar{p}}(\theta_k)] \frac{\partial r_{\bar{p}}(\theta_k)}{\partial p_2} = 0$$

$$\sum_{k=1}^{n} [r_k - r_{\bar{p}}(\theta_k)] \frac{\partial r_{\bar{p}}(\theta_k)}{2p_3} = 0.$$

We find that

$$\frac{\partial r(\theta,p_1,p_2,p_3)}{\partial p_1} = \frac{1}{1 - p_2 \cos (\theta - p_3)}$$

$$\frac{\partial r(\theta,p_1,p_2,p_3)}{\partial p_2} = \frac{p_1 \cos (\theta - p_3)}{[1 - p_2 \cos (\theta - p_3)]^2}$$

and

$$\frac{\partial r(\theta,p_1,p_2,p_3)}{\partial p_2} = \frac{p_1 p_2 \sin (\theta - p_3)}{[1 - p_2 \cos (\theta - p_3)]^2}$$

and the system of nonlinear equations to be solved for $\bar{p}_1$, $\bar{p}_2$, $\bar{p}_3$ is

$$F_1(\bar{p}) = F_1(\bar{p}_1, \bar{p}_2, \bar{p}_3) = 0$$
$$F_2(\bar{p}) = F_2(\bar{p}_1, \bar{p}_2, \bar{p}_3) = 0$$
$$F_3(\bar{p}) = F_3(\bar{p}_1, \bar{p}_2, \bar{p}_3) = 0$$

where

$$F_1(\bar{p}_1, \bar{p}_2, \bar{p}_3) = \sum_{k=1}^{n} \left(r_k - \frac{\bar{p}_1}{1 - \bar{p}_2 \cos(\theta_k - \bar{p}_3)} \right)\left(\frac{1}{1 - \bar{p}_2 \cos(\theta_k - \bar{p}_3)} \right)$$

$$F_2(\bar{p}_1, \bar{p}_2, \bar{p}_3) - \sum_{k=1}^{n} \left(r_k - \frac{\bar{p}_1}{1 - \bar{p}_2 \cos(\theta_k - \bar{p}_3)} \right)\left(\frac{\bar{p}_1 \cos(\theta_k - \bar{p}_3)}{[1 - \bar{p}_2 \cos(\theta_k - \bar{p}_3)]^2} \right)$$

and

$$F_3(\bar{p}_1, \bar{p}_2, \bar{p}_3) = \sum_{k=1}^{n} \left(r_k - \frac{\bar{p}_1}{1 - \bar{p}_2 \cos(\theta_k - \bar{p}_3)} \right)\left(\frac{\bar{p}_1 \bar{p}_2 \sin(\theta_k - \bar{p}_3)}{[1 - \bar{p}_2 \cos(\theta_k - \bar{p}_3)]^2} \right).$$

In order to use Newton's method to solve the system, first we need an initial guess for $\bar{p}$, and we may take

$$\bar{p}^{(0)} = (\bar{p}_1^{(0)}, \bar{p}_2^{(0)}, \bar{p}_3^{(0)}) = \left(\frac{2ME}{E+M}, \frac{M-E}{E+M}, \pi \right).$$

That is, the originally planned elliptical orbit around the sun taking the spaceship from earth to Mars should be a good first approximation to the actual flight course. If it is *not*, we may be in serious trouble, because we may not have enough fuel to make a very large midcourse correction.

Second, to use Newton's method, we need to have the 3×3 Jacobian matrix of partial derivatives

$$F'(p) = \begin{pmatrix} \dfrac{\partial F_1(p)}{\partial p_1} & \dfrac{\partial F_1(p)}{\partial p_2} & \dfrac{\partial F_1(p)}{\partial p_3} \\[2mm] \dfrac{\partial F_2(p)}{\partial p_1} & \dfrac{\partial F_2(p)}{\partial p_2} & \dfrac{\partial F_2(p)}{\partial p_3} \\[2mm] \dfrac{\partial F_3(p)}{\partial p_1} & \dfrac{\partial F_3(p)}{\partial p_2} & \dfrac{\partial F_3(p)}{\partial p_3} \end{pmatrix}.$$

EXERCISE

Write down formulas for the components of the matrix $F'(p)$.

Newton's method for finding $\bar{p}$ is then the iterative method defined by

$$F'(p^{(k)})(p^{(k+1)} - p^{(k)}) = -F(p^{(k)})$$

with

$$p^{(0)} = \begin{pmatrix} \bar{p}_1^{(0)} \\ \bar{p}_2^{(0)} \\ \bar{p}_3^{(0)} \end{pmatrix}.$$

To use the method, we begin with $k = 0$ and evaluate, at each iteration, the vector function

$$F(p^{(k)}) = \begin{pmatrix} F_1(p_1^{(k)}, p_2^{(k)}, p_3^{(k)}) \\ F_2(p_1^{(k)}, p_2^{(k)}, p_3^{(k)}) \\ F_3(p_1^{(k)}, p_2^{(k)}, p_3^{(k)}) \end{pmatrix}$$

and the matrix $F'(p^{(k)})$. Then we solve the linear system

$$F'(p^{(k)})\Delta^{(k)} = -F(p^{(k)})$$

for

$$\Delta^{(k)} = p^{(k+1)} - p^{(k)} = \begin{pmatrix} \Delta_1^{(k)} \\ \Delta_2^{(k)} \\ \Delta_3^{(k)} \end{pmatrix}$$

and put

$$p^{(k+1)} = p^{(k)} + \Delta^{(k)}$$

and can either iterate again or stop with $\bar{p} \approx p^{(k+1)}$.

A reasonable stopping criterion in this application of Newton's method would be to evaluate the distance function

$$D(p^{(k)})$$

at each iteration and stop when $D(p^{(k+1)})$ is no longer decreasing. Thus we could, for instance, iterate until $D(p^{(k+1)}) < .99\ D(p^{(k)})$ *fails* to be true.

Once having found the best-fitting ellipse to the observed positional data, we can assume that the spacecraft is on the elliptical course described by

$$r_{\bar{p}}(\theta) = \frac{\bar{p}_1}{1 - \bar{p}_2 \cos(\theta - \bar{p}_3)}.$$

We can then compute a correction in velocity that will put the craft back on a course which will meet Mars. There are lots of ways to get there and what we want is a correction that involves a minimum expenditure of fuel, or (what amounts to the same thing) a minimum change in the magnitude of the velocity.

The constancy (*conservation*) of energy and of angular momentum that are given in Appendix A as

$$\frac{1}{2}(\dot{x}_1{}^2 + \dot{x}_2{}^2) - \frac{q}{r} = \text{constant}$$

and

$$x_2 \dot{x}_1 - x_1 \dot{x}_2 = \text{constant}$$

can be written conveniently in polar coordinates r, $\theta(x_1 = r \cos \theta, x_2 = r \sin \theta)$ as

$$\frac{1}{2}\{(r\dot{\theta})^2 + \dot{r}^2\} - \frac{g}{r} = \text{constant}$$

and

$$r^2\dot{\theta} = \text{constant}.$$

The "constants" refer to a given orbit.

We can use $r^2\dot{\theta} = r_0^2\dot{\theta}_0$ to determine the *time* required for the spaceship to traverse part of an orbit up to r, θ beginning at r_0, θ_0 at time t_0; namely,

$$r^2\frac{d\theta}{dt} = r_0^2\dot{\theta}_0$$

so

$$t - t_0 = \frac{1}{r_0^2\dot{\theta}_0}\int_{\theta_0}^{\theta} r^2(\theta)d\theta.$$

Suppose we are going to make the midcourse correction at time t_0 when, according to our observations and the determination by least squares of the ellipse $r_{\bar{p}}(\theta)$, the spacecraft will have the position r_0, θ_0. The time beyond t_0 at which the spacecraft will reach its *apogee* (farthest point from the sun) for a *given* orbit $r_p(\theta)$ is the time

$$t_a = t_0 + \frac{1}{r_0^2\dot{\theta}_0}\int_{\theta_0}^{p_3} r_p^2(\theta)d\theta.$$

We want this to be the same as the time when Mars reaches the same position, so we seek an orbit $r_{p*}(\theta)$ such that

$$t_0 + \frac{1}{r_0^2\dot{\theta}_0}\int_{\theta_0}^{p_3^*} r_{p*}^2(\theta)d\theta = \frac{1}{M^2\dot{\theta}_M}\int_{\theta_M}^{p_3^*} M^2 d\theta = \frac{p_3^* - \theta_M}{\dot{\theta}_M}$$

where $\dot{\theta}_M$ is the angular velocity of Mars about the sun (assumed to be $\dot{\theta}_M = 2\pi M/T_M$, where T_M is the *period* of Mars $\approx (1.9)\cdot 365\cdot 24$ hours) and where θ_M is the angular position of Mars (relative to the x_1-axis) at time t_0. Both θ_M and $\dot{\theta}_M$ may be assumed to be known to good approximation (compared to the quantities we seek).

Suppose we are going to use a reasonably "high thrust" (large acceleration) to "correct" the orbit of the spaceship so that the position of the ship will change relatively little during the acceleration (if it does not last very long). (If we were going to use some sort of very low thrust device, which would be left on for many hours or days, then a different set of calculations would be required.)

The orbit we seek,

$$r_{p*}(\theta) = \frac{p_1^*}{1 - p_2^*\cos(\theta - p_3^*)}$$

should reach Mars when $\theta = p_3{}^*$, so we want

$$r_{p^*}(p_3{}^*) = \frac{p_1{}^*}{1 - p_2{}^*} = M.$$

See Figure 5.9.

From the formula for our elliptical orbits

$$r_p(\theta) = \frac{p_1}{1 - p_2 \cos\,(\theta - p_3)}$$

and using the "conservation laws"

$$r^2\dot\theta = r_0{}^2\dot\theta_0$$

$$\frac{1}{2}\,[(r\dot\theta)^2 + \dot r^2] - \frac{g}{r} = \frac{1}{2}\,[(r_0\dot\theta_0)^2 + \dot r_0{}^2] - \frac{g}{r_0}$$

we find, by differentiation and substitutions

$$\text{(using also } \dot r_p(\theta) = (dr_p(\theta)/d\theta)\cdot\dot\theta),$$

that

$$p_1 = \frac{r_0{}^2(r_0\dot\theta_0)^2}{g}$$

and

$$p_2 = \left[1 - \frac{r_0{}^2(r_0\dot\theta_0)^2}{g}\left(\frac{2g}{r_0} - \dot r_0{}^2 - (r_0\dot\theta_0)^2\right) \right]^{1/2}.$$

Call $u = r_0\dot\theta_0$ and $v = \dot r_0$. For the orbit $r_{\bar p}(\theta)$, which we have determined from observations, we have

$$\bar p_1 = \frac{r_0{}^2\bar u^2}{g}$$

and

$$\bar p_2 = \left[1 - \frac{r_0{}^2\bar u^2}{g}\left(\frac{2g}{r_0} - \bar v^2 - \bar u^2\right) \right]^{1/2}$$

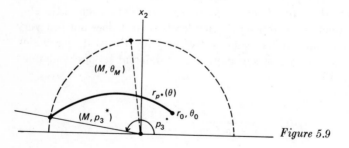

Figure 5.9

and we can solve for $\bar{u}$ and $\bar{v}$ to get

$$\bar{u} = \left(\frac{g\bar{p}_1}{r_0^2}\right)^{1/2}, \qquad \bar{v} = \left(\frac{g\bar{p}_1}{r_0^2} - \frac{1 - \bar{p}_2^2}{\bar{p}_1} - \frac{g\bar{p}_1}{r_0^2}\right)^{1/2}.$$

We seek p_1^*, p_2^*, p_3^* such that

$$\frac{p_1^*}{1 - p_2^*} = M$$

and

$$t_0 + \frac{1}{r_0 u^*}\int_\theta^{p_3^*} \left(\frac{p_1^*}{1 - p_2^* \cos(\theta - p_3^*)}\right)^2 d\theta = \frac{p_3^* - \theta_M}{\theta_M}$$

where

$$u^* = \left(\frac{gp_1^*}{r_0^2}\right)^{1/2} \quad \text{and} \quad v^* = \left(\frac{2g}{r_0} - \frac{1 - p_2^{*2}}{p_1^*} - \frac{gp_1^*}{r_0^2}\right)^{1/2}.$$

Then the desired velocity correction is

$$\Delta(r\dot{\theta}) = u^* - \bar{u}$$
$$\Delta(\dot{r}) = v^* - \bar{v}$$

to be applied at time t_0 when the spaceship is at (r_0, θ_0) with r_0 given by

$$r_0 = \frac{\bar{p}_1}{1 - \bar{p}_2 \cos(\theta_0 - \bar{p}_3)}.$$

We want to minimize $(u^* - \bar{u})^2 + (v^* - \bar{v})^2$ as well. We can get p_1^* from

$$p_1^* = M(1 - p_2^*)$$

once we have p_2^*.

Now we can substitute $M(1 - p_2^*)$ for p_1^* in the expression for v^* to get v^* as a function of p_2^*. Then we can write $(u^* - \bar{u})^2 + (v^* - \bar{v})^2$ as a function of p_2^*, and we can write its derivative with respect to p_2^* and set that equal to zero to minimize $(u^* - \bar{u})^2 + (v^* - \bar{v})^2$, giving an equation for p_2^*. Finally, p_3^* can be determined numerically, if we want it, from the integral expression. From p_1^* and p_2^* we can compute u^* and v^* and obtain $\Delta(r\dot{\theta})$ and $\Delta(\dot{r})$.

The equation to be solved for p_2^* is

$$\frac{d}{dp_2^*}[(u^* - \bar{u})^2 + (v^* - \bar{v})^2] = \frac{d}{dp_2^*}\left[\left(\frac{gM(1 - p_2^*)}{r_0^2}\right)^{1/2} - \bar{u}\right]^2$$
$$+ \frac{d}{dp_2^*}\left\{\left[\frac{2g}{r_0} - \frac{1 + p_2^*}{M} - \frac{gM(1 - p_2^*)}{r_0^2}\right]^{1/2} - \bar{v}\right\}^2 = 0$$

We may use an iterative method for a numerical solution with the starting guess

$$p_2^* \approx \bar{p}_2.$$

EXERCISE

Derive the iteration formula obtained by applying Newton's method to the equation to be solved for $p_2{}^*$.

5.5 Function Spaces and Orthogonality

Just as we say that the components of a vector $(x_1, x_2, \ldots, x_n)$ represent the coordinates of a *point* x in E^n, so the *values* of a discrete function $(f(t_1), f(t_2), \ldots, f(t_n))$ also represent the coordinates of a *point* f in E^n. If f and g are two discrete functions with the same domain, $\{t_1, \ldots, t_n\}$, we can *add* f and g so that $f + g$ is another discrete function (with the same domain) whose value at any t is the sum of the values of f and g at t; thus to get the coordinates of the *point* $f + g$ we add the corresponding coordinates of the points f and g. We can also multiply f by a constant (*scalar*, real number), a, to get another discrete function, af, whose values are a times the values of f so that the coordinates of the *point* af are a times the corresponding of f. Thus the set of *all* discrete real valued functions with a common domain consisting of a finite set of n points forms a *vector space* which is the same as E^n, the set of all n-tuples of real numbers with addition and scalar multiplication defined *componentwise* (or *coordinatewise*).

More generally, the set of *all* real-valued functions with *any* given *common* domain D forms a *function space* $L[D]$ (which will be an infinite dimensional vector space if D is not a finite set) in which the *coordinates* of a *point* f are the values of f. We *add* two functions by adding their values at each element of D. Thus if $h = f + g$, then $h(t) = f(t) + g(t)$ for all t in D. Similarly, if a is a real number, then af is defined to have coordinates $af(t)$ for all t in D.

A *subspace* of a function space $L[D]$ is a set S of points (functions) in $L[D]$ such that whenever f and g are in S, then so is $f + g$ in S and also af in S for any real number a. It follows that any linear combination of elements (points) of a subspace is another element of the subspace. That is, if $f_1, f_2, \ldots, f_K$ are in S and $a_1, a_2, \ldots, a_K$ are real numbers, then $a_1 f_1 + a_2 f_2 + \cdots + a_K f_K$ is in S. Notice that a subspace is itself a function space.

This is an important definition and concept, and we consider now some examples. We consider the space $L[[0,1]]$ of all real-valued functions with common domain $[0,1]$. We can immediately verify that according to our definition of subspace, each of the following is a subspace of $L[[0,1]]$:

1. The continuous functions in $L[[0,1]]$
2. Linear combinations of a chosen finite set of functions in $L[[0,1]]$
3. Polynomials of degree not exceeding some fixed integer N

4. All polynomials
5. Differentiable functions in $L[[0,1]]$
6. All functions in $L[[0,1]]$ that vanish on a chosen subset of $[0,1]$

There are other kinds of subspaces as well, but these will serve to illustrate the richness of the concept. We can have subspaces of subspaces and, in fact, in the examples just given, (3) is a subspace of (4) which is a subspace of (5) which is a subspace of (1).

As a specific example of a subspace of type (6), we can mention the subspace of functions that vanish at every t in $[0,1]$ *except* at a finite set of points $(t_1, t_2, \ldots, t_n)$ in $[0,1]$. We can thus view discrete functions as represented by points in such a subspace.

In Appendix B we mention *inner products* of vectors. If $x = (x_1, x_2, \ldots, x_n)$ and $y = (y_1, y_2, \ldots, y_n)$ are vectors (points) in E^n, then the inner product of x and y is given by

$$(x,y) = x_1 y_1 + x_2 y_2 + \cdots + x_n y_n = \sum_{k=1}^{n} x_k y_k.$$

We say that x is *orthogonal* to y (perpendicular to y) if $(x,y) = 0$. The *norm* or *length* of a vector x in E^n is given by

$$\|x\| = \left(\sum_{k=1}^{n} x_k^2 \right)^{1/2} = (x,x)^{1/2}.$$

If we consider straight lines from the origin through each of the points x and y, then the *angle* θ between the lines satisfies (see Figure 5.10).

$$\cos \theta = \frac{(x,y)}{\|x\| \cdot \|y\|}.$$

Thus when (x,y) is zero, the angle between x and y is a right angle and the lines through x and y are perpendicular.

Some properties of inner products of importance are: symmetry, $(x,y) = (y,x)$: linearity, $(x, au + bv) = a(x,u) + b(x,v)$ for any real numbers a and b and any vectors x, u, and v; and the property that (x,x) is zero only if x is the zero vector.

Recall that the (Euclidean) distance between two points x and y is given by

$$\|x - y\| = (x - y, x - y)^{1/2} = \left(\sum_{k=1}^{n} (x_k - y_k)^2 \right)^{1/2}.$$

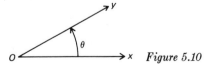

Figure 5.10

If S is a subspace of E^n and x is a point in E^n that is *not* in S, then there will be a point in S which is nearest to x. That is, there will be some point $\bar{y}$ in S such that $\|x - \bar{y}\|$ is the minimum distance between x and any point in S (see Figure 5.11).

We can find $\bar{y}$ by *orthogonal projection* of x on S. If every y in S can be written as a linear combination

$$y = p_1\varphi_1 + p_2\varphi_2 + \cdots + p_M\varphi_M$$

for some fixed set of vectors $\varphi_1, \varphi_2, \ldots, \varphi_M$ and for some coefficients (real numbers) $p_1, p_2, \ldots, p_M$ depending only on y, then

$$\|x - y\| = (x - y, x - y)^{1/2}$$

will be minimum for $y = \bar{y}$ in S if $x - \bar{y}$ is orthogonal to S, that is, to every vector in S. In other words, y can be found from the equations

$$(x - \bar{y}, \varphi_1) = 0$$
$$(x - \bar{y}, \varphi_2) = 0$$
$$\cdot$$
$$\cdot$$
$$\cdot$$
$$(x - \bar{y}, \varphi_M) = 0$$

or, putting $\bar{y} = \bar{p}_1\varphi_1 + \bar{p}_2\varphi_2 + \cdots + \bar{p}_M\varphi_M$, we have, using the linearity of the inner product, the following system of equations to solve for $\bar{p}_1, \ldots, \bar{p}_M$:

$$(\varphi_1,\varphi_1)\bar{p}_1 + (\varphi_1,\varphi_2)\bar{p}_2 + \cdots + (\varphi_1,\varphi_M)\bar{p}_M = (x,\varphi_1)$$
$$(\varphi_2,\varphi_1)\bar{p}_1 + (\varphi_2,\varphi_2)\bar{p}_2 + \cdots + (\varphi_2,\varphi_M)\bar{p}_M = (x,\varphi_2)$$
$$\cdot$$
$$\cdot$$
$$\cdot$$
$$(\varphi_M,\varphi_1)\bar{p}_1 + (\varphi_M,\varphi_2)\bar{p}_2 + \cdots + (\varphi_M,\varphi_M)\bar{p}_M = (x,\varphi_M).$$

Suppose the vectors $\varphi_1, \varphi_2, \ldots, \varphi_M$ form an *orthonormal basis* for S, that is, every y in S can be written as

$$y = p_1\varphi_1 + \cdots + p_M\varphi_M$$

and the inner products (φ_1,φ_j) satisfy

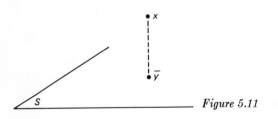

Figure 5.11

$$(\varphi_i, \varphi_j) = \begin{cases} 1 & \text{if } i = j \\ 0 & \text{if } i \neq j. \end{cases}$$

Then the vectors $\varphi_1, \varphi_2, \ldots, \varphi_M$ are *mutually orthogonal, unit* vectors ($\|\varphi_i\| = 1, i = 1,2,\ldots,M$).

In this case, the matrix of coefficients in the linear system of equations for $\bar{p}_1, \bar{p}_2, \ldots, \bar{p}_M$ is the identity matrix and we have the explicit solution

$$\bar{p}_i = (x, \varphi_i) \qquad (i = 1,2,\ldots,M).$$

If the matrix of coefficients (φ_i, φ_j) $(i,j = 1,2,\ldots,M)$ is, in any case, non-singular, then the vectors $\varphi_1, \varphi_2, \ldots, \varphi_M$ are *linearly independent* and there will be a unique solution for $\bar{p}_1, \bar{p}_2, \ldots, \bar{p}_M$.

If we are talking about the set of all discrete functions with domain $\{t_1, t_2, \ldots, t_n\}$, we can view this as the vector space E^n, where each *point* is *one* of the discrete functions, the coordinates of the point f being given by the vector of values of f. So

$$f = [f(t_1), f(t_2), \ldots, f(t_n)].$$

Let $\varphi_1, \varphi_2, \ldots, \varphi_M$ be some chosen set of points (discrete functions with domain $\{t_1, t_2, \ldots, t_n\}$), then the set of all linear combinations, $\{p_1\varphi_1 + p_2\varphi_2 + \cdots + p_M\varphi_M\}$, forms a *subspace* S of E^n. To verify this we need only see that

1. If

$$\varphi^{(1)} = p_1^{(1)}\varphi_1 + p_2^{(1)}\varphi_2 + \cdots + p_M^{(1)}\varphi_M$$

and

$$\varphi^{(2)} = p_1^{(2)}\varphi_1 + p_2^{(2)}\varphi_2 + \cdots + p_M^{(2)}\varphi_M$$

then

$$\varphi^{(1)} + \varphi^{(2)} = (p_1^{(1)} + p_1^{(2)})\varphi_1 + \cdots + (p_M^{(1)} - p_M^{(2)})\varphi_M.$$

So

$$\varphi^{(1)} + \varphi^{(2)} \text{ is in } S.$$

2. If a is a real number, then

$$a\varphi^{(1)} = ap_1^{(1)}\varphi_1 + \cdots + ap_M^{(1)}\varphi_M,$$

so $a\varphi^{(1)}$ is in S.

If f is a point (discrete function) which is *not* in S, we can find the closest point to f that *is* in S by orthogonal projection of f on S by exactly the same set of equations as given in connection with Figure 5.11. The picture is the same (see Figure 5.12).

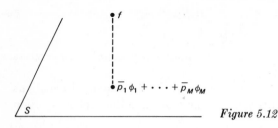

$$\bar{p}_1 \phi_1 + \cdots + \bar{p}_M \phi_M$$

Figure 5.12

If the vectors φ_1, φ_2, ..., φ_M form an orthonormal basis for S, then the coefficients $\bar{p}_1$, ..., $\bar{p}_M$ are given simply by

$$\bar{p}_1 = (f,\varphi_1), \ \bar{p}_2 = (f,\varphi_2), \ldots, \bar{p}_M = (f,\varphi_M).$$

Given any *basis* for S, $(\varphi_1,\varphi_2, \ldots, \varphi_M)$, such that every point y in S can be written as a unique linear combination of the φ's, then *we can construct an orthonormal basis* $\bar{\varphi}_1, \ldots, \bar{\varphi}_M$ as follows (*Gram-Schmidt orthonormalization process*).

Put $\bar{\varphi}_1 = \varphi_1/\|\varphi_1\|$, where $\|\varphi_1\| = (\varphi_1,\varphi_1)^{1/2}$; and recursively determine $\bar{\varphi}_{k+1}$ from $\bar{\varphi}_1, \ldots, \bar{\varphi}_k$ as follows:

$$\varphi_{k+1}^{(1)} = \varphi_{k+1} - \sum_{i=1}^{k} (\varphi_{k+1},\bar{\varphi}_i)\bar{\varphi}_i$$

$$\bar{\varphi}_{k+1} = \left(\frac{1}{\|\varphi_{k+1}^{(1)}\|}\right)\varphi_{k+1}^{(1)} \qquad (k = 1,2, \ldots, M - 1).$$

Each $\bar{\varphi}_{k+1}$ is obtained by subtracting from φ_{k+1} the projection of φ_{k+1} upon the subspace spanned by the previously found $\bar{\varphi}_1, \bar{\varphi}_2, \ldots, \bar{\varphi}_k$ and then making the resulting difference into a unit vector by scalar multiplication by the reciprocal of its length. Notice that for any nonzero vector v we have

$$\left(\frac{1}{\|v\|} v, \frac{1}{\|v\|} v\right) = \frac{1}{\|v\|^2} (v,v) = 1$$

so that $(1/\|v\|) v$ is a unit vector.

As an illustration, consider the subspace S of discrete functions on $\{0, \frac{1}{2}, 1\}$ spanned by $\varphi_1 = (1,1,1)$, $\varphi_2 = (0,\frac{1}{2},1)$. We can view φ_1 and φ_2 as the restrictions to the discrete domain $\{0, \frac{1}{2}, 1\}$ of the continuous functions on $[0,1]$ given by $\varphi_1(t) = 1$, $\varphi_2(t) = t$. So S is the subspace of all discrete functions y on $\{0, \frac{1}{2}, 1\}$ of the form $y = p_1\varphi_1 + p_2\varphi_2$.

Notice that the inner product $(\varphi_1,\varphi_2) = 1.0 + 1 \cdot \frac{1}{2} + 1 \cdot 1 = 1\frac{1}{2}$ does not vanish, so φ_1 and φ_2 are *not* already orthogonal. However, using the Gram-Schmidt process we can find

$$\bar{\varphi}_1 = \frac{\varphi_1}{\|\varphi_1\|}$$

and

$$\bar{\varphi}_2 = \frac{\varphi_2 - (\varphi_2,\bar{\varphi}_1)\bar{\varphi}_1}{\|\varphi_2 - (\varphi_2,\bar{\varphi}_1)\bar{\varphi}_1\|},$$

so that $(\bar{\varphi}_1,\bar{\varphi}_2) = 0$. We have $\|\varphi_1\| = (\varphi_1,\varphi_1)^{1/2} = \sqrt{3}$, so

$$\bar{\varphi}_1 = \left(\frac{1}{\sqrt{3}},\frac{1}{\sqrt{3}},\frac{1}{\sqrt{3}}\right).$$

Now $(\varphi_2,\bar{\varphi}_1) = 3/2\sqrt{3} = \sqrt{3}/2$, so

$$\varphi_2 - (\varphi_2,\bar{\varphi}_1)\bar{\varphi}_1 = (0,\tfrac{1}{2},1) - \frac{\sqrt{3}}{2}\left(\frac{1}{\sqrt{3}},\frac{1}{\sqrt{3}},\frac{1}{\sqrt{3}}\right) = (-\tfrac{1}{2},0,\tfrac{1}{2})$$

and

$$\|\varphi_2 - (\varphi_2,\bar{\varphi}_1)\bar{\varphi}_1\| = \sqrt{\tfrac{1}{4} + 0 + \tfrac{1}{4}} = \frac{1}{\sqrt{2}};$$

therefore

$$\bar{\varphi}_2 = \left(-\frac{1}{\sqrt{2}},0,\frac{1}{\sqrt{2}}\right).$$

We can check that $(\bar{\varphi}_1,\bar{\varphi}_2) = 0$.

Using $\bar{\varphi}_1$, $\bar{\varphi}_2$ as a basis for S, we find the closest point in S to any discrete function f on $\{0, \tfrac{1}{2}, 1\}$, easily, by orthogonal projection of f on S. We have

$$\bar{p}_1\bar{\varphi}_1 + \bar{p}_2\bar{\varphi}_2$$

as the closest point in S to f if $\bar{p}_1$, $\bar{p}_2$ are chosen as $\bar{p}_1 = (f,\bar{\varphi}_1)$ and $\bar{p}_2 = (f,\bar{\varphi}_2)$. For instance, if f is the restriction to $\{0, \tfrac{1}{2}, 1\}$ of $f(t) = t^2$, then

$$f = (0,\tfrac{1}{4},1)$$

and $\bar{p}_1 = (f,\bar{\varphi}_1) = 1/(4\sqrt{3}) + 1/(\sqrt{3}) = 5/4\sqrt{3}$ and $\bar{p}_2 = (f,\bar{\varphi}_2) = 1/\sqrt{2}$ so $(5/4\sqrt{3})\bar{\varphi}_1 + (1/\sqrt{2})\bar{\varphi}_2 = [-(1/12)(5/12),(11/12)]$ is the closest point in S to f.

The discrete functions $\bar{\varphi}_1$ and $\bar{\varphi}_2$, which we found in this example, can also be seen to be the restrictions to $\{0,\tfrac{1}{2},1\}$ of the continuous functions $\bar{\varphi}_1(t) = (1/\sqrt{3})$, $\bar{\varphi}_2(t) = \sqrt{2}[t - (\tfrac{1}{2})]$ and the closest point in S to f is the restriction to $(0,\tfrac{1}{2},1)$ of the continuous function

$$\frac{5}{4\sqrt{3}}\,\bar{\varphi}_1(t) + \frac{1}{\sqrt{2}}\,\bar{\varphi}_2(t) = \frac{5}{12} + \left(t - \frac{1}{2}\right) = t - \frac{1}{12}.$$

We turn our attention now to the subspace of *continuous* functions on $[0,1]$. Recall that a real-valued continuous function f with domain $[0,1]$ is *locally* constant so that if we take enough points $t_1, t_2, \ldots, t_n$ in $[0,1]$ the value of f at any t in $[0,1]$ will be as close as we please to the value of f at any

t_i nearby to t. For instance, let $t_i = i/n$, $(i = 1,2,3, \ldots, n)$; then for a given continuous function f and a given integer N we will have

$$|f(t) - f(t_i)| < \frac{1}{N}, \qquad t_i - \frac{i}{n} \leq t \leq t_i$$

for all $i = 1, 2, \ldots, n$ *if n is sufficiently large.*

We can approximate a continuous function f on [0,1] by a *step function* (piecewise constant function) in a standard way as follows: denote by $s_n f$ the step-function approximation to f (using n steps) given by

$$s_n f(t) = \begin{cases} f\!\left(\dfrac{1}{n}\right) & \text{for } 0 \leq t \leq \dfrac{1}{n}; \\[2ex] f\!\left(\dfrac{i}{n}\right) & \text{for } \dfrac{i-1}{n} < t \leq \dfrac{i}{n}, \end{cases} \qquad \text{with } i = 2, 3, \ldots, n.$$

For a given continuous function f on [0,1] we will have

$$\lim_{n \to \infty} s_n f(t) = f(t)$$

for every t in [0,1].

Consider for a moment the problem of minimizing the distance between a given *discrete* function f with domain $\{0, 1/n, 2/n, \ldots, 1\}$ and a subspace S of discrete functions with the same domain. We seek φ in S such that

$$D(\varphi) = \left(\sum_{i=0}^{n} \left(f\!\left(\frac{i}{n}\right) - \varphi\!\left(\frac{i}{n}\right) \right)^2 \right)^{1/2}$$

is minimum. But if we minimize $D(\varphi)$, we will also minimize, at the same time, any positive constant times $D(\varphi)$, for instance $(1/\sqrt{n})D(\varphi)$.

Let f and g be any two continuous functions on [0,1].

Define the *n-step-distance* between f and g as

$$D_n(f,g) = \left(\frac{1}{n} \sum_{i=1}^{n} \left(f\!\left(\frac{i}{n}\right) - g\!\left(\frac{i}{n}\right) \right)^2 \right)^{1/2}.$$

Notice that

$$D_n(f,g) = \left(\int_0^1 (s_n f(t) - s_n g(t))^2 dt \right)^{1/2}$$

because $(s_n f(t) - s_n g(t))^2$ is constant in the intervals $(i-1)/n < t < i/n$ $(i = 1, 2, \ldots, n)$.

Since

$$\lim_{n \to \infty} s_n f(t) = f(t)$$

and

$$\lim_{n \to \infty} s_n g(t) = g(t),$$

it follows that

$$\lim_{n\to\infty} D_n(f,g) = \left(\int_0^1 (f(t) - g(t))^2 dt\right)^{1/2}$$

We can conclude from all this discussion that the effect of taking more and more points (or steps) in the domain of a discrete (or step) function approximation to a continuous function on [0,1] — as far as the problem of finding a *closest point* in some subspace of approximating functions is concerned — is to come closer and closer to dealing with the definition of *distance between two continuous functions given by*

$$D(f,g) = \left(\int_0^1 (f(t) - g(t))^2\right)^{1/2}.$$

We can arrive at such a definition by yet another route as will now be shown.

For finite dimensional vectors $f = (f_1, f_2, \ldots, f_n)$ and $g = (g_1, g_2, \ldots, g_n)$ we define the *inner product* as $(f,g) = \sum_{i=1}^{n} f_i g_i$.

For continuous functions on [0,1] *we define the* inner product *as*

$$(f,g) = \int_0^1 f(t)g(t)dt = \lim_{n\to\infty} \left\{\frac{1}{n} \sum_{i=1}^{n} f\left(\frac{i}{n}\right)g\left(\frac{i}{n}\right)\right\}.$$

We can check that this inner product, so defined, has the properties of symmetry: $(f,g) = (g,f)$, and linearity:

$$(f, ag + bh) = a(f,g) + b(f,h).$$

EXERCISE

Give an argument (proof?) to show that, for continuous f on [0,1], $(f,f) = 0$ implies $f(t) \equiv 0$ for all t in [0,1].

Using the inner product for continuous functions we can define a *norm* (called the "L_2-norm") of a continuous function on [0,1] as

$$\|f\| = (f,f)^{1/2} = \left\{\int_0^1 f^2(t)dt\right\}^{1/2}.$$

The distance between two continuous functions f and g, using this norm, is given by

$$\|f - g\| = (f - g, f - g)^{1/2} = \left\{\int_0^1 [f(t) - g(t)]^2 dt\right\}^{1/2}.$$

This is the same as the *distance* previously derived and denoted by $D(f,g)$. We will use the notation $\|f - g\|$ most of the time.

With the definition of inner product given for continuous functions, we can now say that a continuous function f is orthogonal on $[0,1]$ to a continuous function g if

$$(f,g) = \int_0^1 f(t)g(t)dt = 0.$$

EXERCISES

1. Show that the second-degree polynomial $p(t) = t^2 - t + 1/6$ is orthogonal on $[0,1]$ to every first-degree (linear) polynomial $q(t) = a + bt$.
2. What are *all* the second-degree polynomials that are orthogonal on $[0,1]$ to $q(t) = a + bt$ for every a and b?

5.6   Optimal Approximation and Projections

We can use the concepts and notation of the previous section to show how to obtain *optimal* (or *best*) linear combinations of various chosen forms of approximating continuous functions. The idea is simply to determine the *orthogonal projection* of the continuous function f to be approximated on the subspace of approximating continuous functions. In terms of inner products, the formulas and methods will be nearly identical to those given in the previous section for discrete functions.

Consider now the vector space C of real-valued functions that are continuous on $[0,1]$ and adopt the notation of the previous section for inner products, norms and distances in C.

If S is a subspace of C, then we say that *g in C is orthogonal to the subspace S if g is orthogonal to every function in S.*

We will consider only subspaces that have a finite basis $\{\varphi_1, \varphi_2, \ldots, \varphi_M\}$ of functions such that every y in S is a linear combination of $\varphi_1, \varphi_2, \ldots, \varphi_M$.

Given an f in C, to find the optimal approximation $\bar{y}$ in S, we seek the nearest point in S to f. We want to minimize the distance

$$\| f - y \| = \left(\int_0^1 (f(t) - y(t))^2 \right)^{1/2}$$

for y in S. This is again *optimal* approximation in the sense of *least squares*. We can do this by finding the point $\bar{y}$ in S for which $g = f - \bar{y}$ is orthogonal to the subspace S, just as we did in the previous section for discrete functions (see Figures 5.12 and 5.13).

If S has the basis $\{\varphi_1, \varphi_2, \ldots, \varphi_M\}$, then $\bar{y}$ must satisfy

$$(f - \bar{y}, \varphi_i) = 0 \qquad (i = 1, 2, \ldots, M).$$

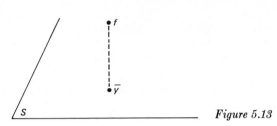

Figure 5.13

Since $\bar{y}$ has the form

$$\bar{y} = \bar{p}_1\varphi_1 + \bar{p}_2\varphi_2 + \cdots + \bar{p}_M\varphi_M$$

we obtain a linear algebraic system to solve for $\bar{p}_1, \bar{p}_2, \ldots, \bar{p}_M$:

$$(f - \bar{p}_1\varphi_1 - \bar{p}_2\varphi_2 - \cdots - \bar{p}_M\varphi_M, \varphi_i) = 0 \qquad (i = 1, 2, \ldots, M)$$

or, in other words (using the linearity of the inner product),

$$(\varphi_1,\varphi_1)\bar{p}_1 + (\varphi_1,\varphi_2)\bar{p}_2 + \cdots + (\varphi_1,\varphi_M)\bar{p}_M = (f,\varphi_1)$$
$$(\varphi_2,\varphi_1)\bar{p}_1 + (\varphi_2,\varphi_2)\bar{p}_2 + \cdots + (\varphi_2,\varphi_M)\bar{p}_M = (f,\varphi_2)$$
$$\cdot$$
$$\cdot$$
$$\cdot$$
$$(\varphi_M,\varphi_1)\bar{p}_1 + (\varphi_M,\varphi_2)\bar{p}_2 + \cdots + (\varphi_M,\varphi_M)\bar{p}_M = (f,\varphi_M).$$

If $\{\varphi_1, \varphi_2, \ldots, \varphi_M\}$ is an orthonormal basis, then the solution of the system is given by $\bar{p}_i = (f,\varphi_i)$ $(i = 1, 2, \ldots, n)$; otherwise we can solve it numerically.

If the vectors $\varphi_1, \ldots, \varphi_M$ are *nearly* orthonormal (that is, if the matrix of inner products (φ_i,φ_j) is *nearly* the identity), then the system can be solved efficiently by iteration (see Chapter 4). On the other hand, it can happen for a particular basis that the linear system to be solved is ill-conditioned. Clearly, such a basis is a poor choice from the numerical point of view.

Since the inner products here are all integrals, numerical quadrature formulas may be required to evaluate them (see Chapter 6).

We can now illustrate optimal approximation by orthogonal projection on various subspaces of particular interest or usefulness. Polynomials are useful because they are easy to evaluate and can give good approximation to almost any continuous function at least over a limited range. In fact, there is a mathematical theorem due to Weierstrass that *any* real-valued continuous function on [0,1], for instance, can be approximated as closely as one pleases by *some* polynomial (of high enough degree). This is *not* to say that there are no difficulties in *finding* such a polynomial or that polynomials are always to be desired.

We will conclude this section with an illustration of optimal approximation (in the sense of least squares) by orthogonal projection of a continuous function on a subspace of approximating functions.

Suppose we want to approximate $f(t) = e^{-t^2}$ for t in $[0,1]$ by a polynomial of degree 4.

Let S be the subspace of continuous functions consisting of polynomials of degree 4 or less. If y is in S, then

$$y = a_0 + a_1 t + a_2 t^2 + a_3 t^3 + a_4 t^4.$$

The optimal approximation in S to f is obtained by solving the system

$$(f - \bar{y}, \varphi_i) = 0 \qquad (i = 0,1,2,3,4)$$

where $f(t) = e^{-t^2}$, $\varphi_0(t) = 1$, $\varphi_1(t) = t, \ldots, \varphi_4(t) = t^4$ and $\bar{y}(t) = \bar{a}_0 + \bar{a}_1 t + \cdots + \bar{a}_4 t^4 = \bar{a}_0 \varphi_0(t) + \cdots + \bar{a}_4 \varphi_4(t)$ with $\bar{a}_0, \ldots, \bar{a}_4$ to be found. The system of equations for $\bar{a}_0, \ldots, \bar{a}_4$ is

$$(\varphi_0, \varphi_0)\bar{a}_0 + (\varphi_0, \varphi_1)\bar{a}_1 + \cdots + (\varphi_0, \varphi_4)\bar{a}_4 = (f, \varphi_0)$$

$$\cdot$$
$$\cdot$$
$$\cdot$$

$$(\varphi_4, \varphi_0)\bar{a}_0 + \cdots \cdots \cdots \cdots + (\varphi_4, \varphi_4)\bar{a}_4 = (f, \varphi_4)$$

where

$$(\varphi_i, \varphi_j) = \int_0^1 t^i t^j dt = \frac{1}{i + j + 1} \qquad (i,j = 0,1,2,3,4)$$

and

$$(f, \varphi_i) = \int_0^1 e^{-t^2} t^i dt \qquad (i = 0,1,2,3,4).$$

EXERCISES

1. Using a computer, solve the above system of equations for $\bar{a}_0$, $\bar{a}_1$, $\bar{a}_2$, $\bar{a}_3$, $\bar{a}_4$ and plot both the resulting polynomial

    $$\bar{y}(t) = \bar{a}_0 + \bar{a}_1 t + \bar{a}_2 t^2 + \bar{a}_3 t^3 + \bar{a}_4 t^4$$

    and e^{-t^2} for $t = 0, 0.1, 0.2, \ldots, 1.0$.

2. Use the Gram-Schmidt process to find an orthonormal basis for the subspace S of polynomials of degree 4 or less on $[0,1]$. Use this basis to find the orthogonal projection of $f(t) = e^{-t^2}$ on S. Compare results and effort with Exercise 1.

The matrix of inner products (of powers of t) occurring above is a famous example of an ill-conditioned matrix (a *finite segment of the Hilbert* matrix). All the more reason to use an orthonormal basis instead.

*3. Estimate the condition number of the $n \times n$ Hilbert segment as a function of n.

5.7 Harmonic Analysis, Finite Fourier Series

Many wave forms arising from oscillatory phenomena can be fitted to good approximation by *finite Fourier series*, that is, by linear combinations of sines and cosines with frequencies that are multiples (*harmonics*) of an appropriate fundamental frequency related to the period of the given wave form.

Suppose that f is a continuous function defined on the whole real line; we say that f is *periodic with period* T, if $f(t + T) = f(t)$ for every t.

The graph of such an f might look like that shown in Figure 5.14, for instance. We can pick any interval of length T on the t-axis, for instance, $[0,T]$, and consider the function f there. If we know the values $f(t)$ for all t in $[0,T]$, then we know them for all t, because we can reduce a given t to some t' in the interval $[0,T]$ using $t \equiv t'$ (modulo T). Any real number t will lie in one of the intervals $[0,T] + kT$ ($k = \pm 1, \pm 2, \ldots$) and so $t' = t - kT$ will be in $[0,T]$ and $f(t') = f(t)$.

If we approximate f over the interval $[0,T]$ by a linear combination of other functions *which are also periodic with period* T, then we will have an approximation to f for *all* t.

The trigonometric functions $\varphi_1, \varphi_2, \ldots$ given by

$$\varphi_k(t) = \sin \frac{2\pi kt}{T} \qquad (k = 1,2,3,\ldots)$$

are all periodic with period T. That is,

$$\varphi_k(t + T) = \sin \frac{2\pi k(t + T)}{T} = \sin \frac{2\pi kt}{T} = \varphi_k(t).$$

For a given $k > 1$, the function φ_k has the shorter periods $T/k, 2T/k, \ldots,$ $(k - 1)T/k$ as well. The *frequency* of a periodic function is defined as the reciprocal of its *shortest* period. So the frequency of φ_k is $k(1/T)$.

Any *constant* function, for instance, φ_0 given by $\varphi_0(t) \equiv 1$ for all t, is periodic with period T also (in fact for *any* T) because $\varphi_0(t + T) = \varphi_0(t) = 1$ for all t.

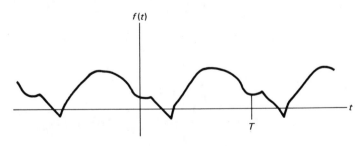

Figure 5.14

The trigonometric functions ψ_1, ψ_2, ... given by

$$\psi_k(t) = \cos \frac{2\pi kt}{T} \qquad (k = 1,2,3, \ldots)$$

are also periodic with period T.

For any positive integer K the functions φ_0, φ_1, φ_2, ..., φ_K, ψ_1, ψ_2, ..., ψ_K are mutually orthogonal. That is, let Φ_K be the set of functions

$$\Phi_K = (\varphi_0, \varphi_1, \ldots, \varphi_K, \psi_1, \psi_2, \ldots, \varphi_K)$$

then

$$(y_1, y_2) = \int_0^T y_1(t)y_2(t)dt = 0$$

for any two different functions y_1 and y_2 in Φ_K.

We can verify this as follows. First,

$$(\varphi_0, \varphi_k) = \int_0^T \sin \frac{2\pi kt}{T} \, dt = \left\{ -\frac{T}{2\pi k} \cos \frac{2\pi kt}{T} \right\}\Big|_0^T = 0, \quad (k = 1,2, \ldots, K)$$

and

$$(\varphi_0, \psi_k) = \int_0^T \cos \frac{2\pi kt}{T} \, dt = \left\{ +\frac{T}{2\pi k} \sin \frac{2\pi kt}{T} \right\}\Big|_0^T = 0, \quad (k = 1,2, \ldots, K).$$

Next,

$$\begin{aligned}
(\varphi_k, \varphi_j) &= \int_0^T \sin \frac{2\pi kt}{T} \sin \frac{2\pi jt}{T} \, dt \\
&= \int_0^T \frac{1}{2} \left\{ \cos \frac{2\pi(k-j)t}{T} - \cos \frac{2\pi(k+j)t}{T} \right\} dt \\
&= 0 \quad \text{for } k \neq j \qquad (k,j = 1,2, \ldots, K)
\end{aligned}$$

and

$$\begin{aligned}
(\varphi_k, \psi_j) &= \int_0^T \sin \frac{2\pi kt}{T} \cos \frac{2\pi kt}{T} \, dt \\
&= \int_0^T \frac{1}{2} \left\{ \sin \frac{2\pi(k-j)t}{T} + \sin \frac{2\pi(k+j)t}{T} \right\} dt \\
&= 0 \quad \text{for } k, j = 1, 2, \ldots, K
\end{aligned}$$

and

$$\begin{aligned}
(\psi_k, \psi_j) &= \int_0^T \cos \frac{2\pi kt}{T} \cos \frac{2\pi jt}{T} \, dt \\
&= \int \frac{1}{2} \left\{ \cos \frac{2\pi(k-j)t}{T} + \cos \frac{2\pi(k+j)t}{T} \right\} dt \\
&= 0 \quad \text{for } k \neq j \qquad (k,j = 1,2, \ldots, K).
\end{aligned}$$

Now the set of all functions that are continuous on the whole real line and are periodic with period T forms a vector space P_T of functions since:

1. $f(t + T) = f(t)$ for all t and
   $g(t + T) = g(t)$ for all t
   implies $f + g$ given by $(f + g)(t) = f(t) + g(t)$ is also periodic with period T:

   $$(f + g)(t + T) = f(t + T) + g(t + T) = f(t) + g(t)$$
   $$= (f + g(t);$$

   and
2. $(af)(t + t) = af(t + t) = af(t) = (af)(t)$ for any constant a.

The set of linear combinations of functions in Φ_K forms a subspace S_K of P_T. We verify this as follows.

An element y of S_K is of the form $y = a_0\varphi_0 + \cdots + a_K\varphi_K + b_1\psi_1 + \cdots + b_K\psi_K$ for some choice of the numbers $a_0, a_1, \ldots, a_K, b_1, \ldots, b_K$.

If $y^{(1)}$ and $y^{(2)}$ are in S_K, then

$$y^{(1)} = a_0^{(1)}\varphi_0 + \cdots + b_K^{(1)}\psi_K$$

and

$$y^{(2)} = a_0^{(2)}\varphi_0 + \cdots + b_K^{(1)}\psi_K$$

and

$$y^{(1)} + y^{(2)} = (a_0^{(1)} + a_0^{(2)})\varphi_0 + \cdots + (b_K^{(1)} + b_K^{(2)})\psi_K$$

is also in S_K. Furthermore

$$ay^{(1)} = (aa_0^{(1)})\varphi_0 + \cdots + (ab_K^{(1)})\psi_K$$

is in S_K as well.

An element y of S_K is called a *finite Fourier series*, and y has the form of a constant plus a linear combination of sines and cosines

$$y = a_0 + \sum_{k=1}^{K} a_k \sin \frac{2\pi kt}{T} + \sum_{k=1}^{K} b_k \cos \frac{2\pi kt}{T}.$$

The functions in Φ_K already form an orthogonal basis for S_K and we can *normalize* them to get an *orthonormal basis* for S_K as follows:

(1) $$\|\varphi_0\| = (\varphi_0,\varphi_0)^{1/2} = \left\{ \int_0^T 1^2 \cdot dt \right\}^{1/2} = T^{1/2},$$

so call $\bar{\varphi}_0 = \varphi_0/T^{1/2}$, then $\|\bar{\varphi}_0\| = 1$;

(2) $$\|\varphi_k\| = \left\{ \int_0^T \sin^2 \frac{2\pi kt}{T} dt \right\}^{1/2} = \left\{ \int_0^T \frac{1 - \cos \frac{4\pi kt}{T}}{2} dt \right\}^{1/2}$$
$$= \left(\frac{T}{2} \right)^{1/2} \quad (k = 1, 2, \ldots, K)$$

so call $\bar{\varphi}_k = \varphi_k/(T/2)^{1/2}$, then $\|\bar{\varphi}_k\| = 1$;

$$(3) \qquad \|\psi_k\| = \left\{ \int_0^T \cos^2 \frac{2\pi kt}{T}\, dt \right\}^{1/2} = \left\{ \int_0^T \frac{1 + \cos \dfrac{4\pi kt}{T}}{2}\, dt \right\}^{1/2}$$

$$= \left(\frac{T}{2} \right)^{1/2} \qquad (k = 1,2, \ldots, K)$$

so call $\bar{\psi}_k = \psi_k/(T/2)^{1/2}$, then $\|\bar{\psi}_k\| = 1$.

The set of functions $\bar{\Phi}_K = (\bar{\varphi}_0, \bar{\varphi}_1, \ldots, \bar{\varphi}_K, \bar{\psi}_1, \ldots, \bar{\psi}_K)$ given by

$$\bar{\varphi}_0(t) \equiv \frac{1}{T^{1/2}}$$

$$\bar{\varphi}_k(t) = \left(\frac{2}{T} \right)^{1/2} \sin \frac{2\pi kt}{T} \qquad (k = 1,2, \ldots, K),$$

$$\bar{\psi}_k(t) = \left(\frac{2}{T} \right)^{1/2} \cos \frac{2\pi kt}{T} \qquad (k = 1,2, \ldots, K)$$

forms an orthonormal basis for S_K and we can find the best approximation $\bar{y}$ in S_K (in the sense of least squares, that is, using the L_2-norm) to a given periodic function f in P_T (with period T) as follows.

The orthogonal projection of f on S_K is

$$\bar{y} = \bar{a}_0 \bar{\varphi}_0 + \bar{a}_1 \bar{\varphi}_1 + \cdots + \bar{a}_K \varphi_K + b_1 \bar{\psi}_1 + \cdots + b_K \bar{\psi}_K$$

where

$$\bar{a}_0 = (f, \bar{\varphi}_0) = \int_0^T f(t) \bar{\varphi}_0(t) dt$$

$$\bar{a}_k = (f, \bar{\varphi}_k) = \left(\frac{2}{T} \right)^{1/2} \int_0^T f(t) \sin \frac{2\pi kt}{T}\, dt \qquad (k = 1,2, \ldots, K)$$

$$b_k = (f, \bar{\psi}_k) = \left(\frac{2}{T} \right)^{1/2} \int_0^T f(t) \cos \frac{2\pi kt}{T}\, dt \qquad (k = 1,2, \ldots, K).$$

This $\bar{y}$ is called the finite *Fourier series for f*.

Suppose we are only given a finite set of values of a function f which is assumed to be continuous and periodic with period T. Say we are given the data $[f(t_1), f(t_2), \ldots, f(t_n)]$ as a set of values of f at points $t_1, t_2, \ldots, t_n$ with $0 \le t_1 < t_2 < \cdots < t_n < T$.

In order to find a continuous finite Fourier-series approximation, say an element of S_K, to f we would have to at least *approximately* evaluate the integrals occurring in the expressions for the coefficients $\bar{a}_0, \bar{a}_1, \ldots, \bar{a}_K$, $b_1, \ldots, b_K$, with a quadrature formula that uses just the values of f at the given data points.

An alternative and more direct approach to such an approximation would be to find the orthogonal projection of the *discrete* function f, given by the data, on the subspace $S_K[T_n]$ of discrete functions on $T_n = (t_1, t_2, \ldots, t_n)$ consisting of linear combinations of the discrete functions $\bar{\varphi}_0 \mid_{T_n}, \ldots, \bar{\psi}_K \mid_{T_n}$ obtained by restricting the functions in $\bar{\Phi}_K$ to $T_n = (t_1, t_2, \ldots, t_n)$.

We can now derive formulas for the set of coefficients $\bar{a}_0, \bar{a}_1, \ldots, \bar{a}_K,$ $\bar{b}_1, \ldots, \bar{b}_K$ which minimizes the distance $\| f - \bar{y} \mid_{T_n} \|$ from a discrete function f on T_n to the subspace $S_K[T_n]$, with

$$\| f - \bar{y} \mid_{T_n} \| = \left\{ \sum_{i=1}^{n} [f(t_i) - \bar{y}(t_i)]^2 \right\}^{1/2}$$

where $\bar{y} = \bar{a}_0 \bar{\varphi}_0 + \bar{a}_1 \bar{\varphi}_1 + \cdots + \bar{a}_K \bar{\varphi}_K + \bar{b}_1 \bar{\psi}_1 + \cdots + \bar{b}_K \bar{\psi}_K$.

We want $f - \bar{y} \mid_{T_n}$ to be orthogonal to each of the discrete functions $\bar{\varphi}_0 \mid_{T_n}, \ldots, \bar{\psi}_K \mid_{T_n}$. The resulting equations for $\bar{a}_0, \bar{a}_1, \ldots, \bar{b}_K$ are

$$(\bar{\varphi}_0 \mid_{T_n}, \bar{\varphi}_0 \mid_{T_n}) \bar{a}_0 + \cdots + (\bar{\varphi}_0 \mid_{T_n}, \bar{\psi}_K \mid_{T_n}) \bar{b}_K = (\bar{\varphi}_0 \mid_{T_n}, f)$$

$$\vdots$$

$$(\bar{\psi}_K \mid_{T_n}, \bar{\varphi}_0 \mid_{T_n}) \bar{a}_0 + \cdots + (\bar{\psi}_K \mid_{T_n}, \bar{\psi}_K \mid_{T_n}) \bar{b}_K = (\bar{\psi}_K \mid_{T_n}, f).$$

Clearly we must have $n \geq 2K + 1$ for a unique solution since the dimension of the subspace $S_K[T_n]$ is $2K + 1$ and the dimension of the vector space of *all* discrete functions on T_n is n. If $2K + 1 = n$, presumably f is *in* $S_K[T_n]$ and we should get $\| f - \bar{y} \mid_{T_n} \| = 0$; otherwise, if $n > 2K + 1$, we will have $\| f - \bar{y} \mid_{T_n} \| > 0$, if f is not in $S_K[T_n]$.

For a specific numerical example, suppose we choose the set T_n, for $n = 5$, as $T_5 = \{0, 0.6, 1.3, 2.2, 2.6\}$ and suppose we have data in the form of a discrete function f on T_5, say

$$f = (0.4, 0.5, -0.4, 1.0, 1.2).$$

In tabular form the data can be written

i	t_i	$f(t_i)$
1	0	0.4
2	0.6	0.5
3	1.3	−0.4
4	2.2	1.0
5	2.6	1.2

Assume that the data represent partial information about a function *assumed to be* periodic with period $T = 3.5$.

Let us put $K = 1$ and seek the best *approximation in* $S_1[T_5]$ to the given data. That is, we seek coefficients $\bar{a}_0, \bar{a}_1, \bar{b}_1$ such that we get an approximation of the form

$$\bar{y}(t) = \bar{a}_0\bar{\varphi}_0(t) + \bar{a}_1\bar{\varphi}_1(t) + \bar{b}_1\bar{\psi}_1(t)$$

to $f(t)$ for $0 \le t \le 3.5$, which is *best in the sense that*

$$\left\{\sum_{i=1}^{n} (f(t_i) - \bar{y}(t_i))^2\right\}^{1/2}$$

is minimum *among all choices* of $\bar{a}_0$, $\bar{a}_1$, $\bar{b}_1$.

Since $T = 3.5$ here, we have (carrying two decimal figures):

$$\bar{\varphi}_0(t) = \frac{1}{(3.5)^{1/2}} = .53$$

$$\bar{\varphi}_1(t) = \left(\frac{2}{3.5}\right)^{1/2} \sin\frac{2\pi}{3.5}t = .75 \sin (1.8)t$$

$$\bar{\psi}_1(t) = \left(\frac{2}{3.5}\right)^{1/2} \cos\frac{2\pi}{3.5}t = .75 \cos (1.8)t.$$

We can show the discrete functions $\bar{\varphi}_0|_{T_5}$, $\bar{\varphi}_1|_{T_5}$, and $\bar{\psi}_1|_{T_5}$ in tabular form as follows (computed by hand using sine-cosine tables to get the values of $\bar{\varphi}_1$ and $\bar{\psi}_1$ to two decimal digits):

i	t_i	$\bar{\varphi}_0\|_{T_5}(t_i) \equiv .53$
1	0	.53
2	0.6	.53
3	1.3	.53
4	2.2	.53
5	2.6	.53

i	t_i	$\bar{\varphi}_1\|_{T_5}(t_i) = .75 \sin (1.8)t_i$
1	0	0
2	0.6	.75 (.88) = .66
3	1.3	.75 (.72) = .54
4	2.2	.75 (−.73) = −.55
5	2.6	.75 (−1.0) = −.75

i	t_i	$\psi_1\|_{T_5}(t_i) = .75 \cos (1.8)t_i$
1	0	.75
2	0.6	.75 (.47) = .35
3	1.3	.75 (−.70) = −.53
4	2.2	.75 (−.68) = −.51
5	2.6	.75 (−.03) = −.02

Using the tables just computed, we can form the various inner products of the form

$$(u|_{T_5}, v|_{T_5}) = \sum_{i=1}^{5} u(t_i) \cdot v(t_i)$$

which occur as coefficients in the system of equations to be solved for $\bar{a}_0, \bar{a}_1, \bar{b}_1$.

We have (to two decimals)

$(\bar{\varphi}_0|_{T_5}, \bar{\varphi}_0|_{T_5}) = 5(.53)^2 = 1.4$

$(\bar{\varphi}_0|_{T_5}, \bar{\varphi}_1|_{T_5}) = .53(0 + .66 + .54 - .55 - .75) = -0.05$

$(\bar{\varphi}_0|_{T_5}, \bar{\psi}_1|_{T_5}) = .53(.75 + .35 - .53 - .51 - .02) = 0.02$

$(\bar{\varphi}_0|_{T_5}, f) = .53(0.4 + 0.5 - 0.4 + 1.0 + 1.2) = 1.4$

$(\bar{\varphi}_1|_{T_5}, \bar{\varphi}_1|_{T_5}) = (.66^2 + .54^2 + .55^2 + .75^2) = 1.6$

$(\bar{\varphi}_1|_{T_5}, \bar{\psi}_1|_{T_5}) = (.66)(.35) + (.54)(-.53) + (-.55)(-.51) + (-.75)(-.02)$
$= .24$

$(\bar{\varphi}_1|_{T_5}, f) = (.66)(0.5) + (.54)(-0.4) + (-.55)(1.0) + (-.75)(1.2)$
$= -1.3$

$(\bar{\psi}_1|_{T_5}, \bar{\psi}_1|_{T_5}) = (.75^2 + .35^2 + .53^2 + .51^2 + .02^2) = 1.2$

$(\bar{\psi}_1|_{T_5}, f) = (.75)(0.4) + (3.5)(0.5) + (-.53)(-0.4) + (-.51)(1.0)$
$+ (-.02)(1.2) = 0.16.$

Since the inner products are symmetric, we can now write down the numerical values (approximately) of the coefficients in the system of equations to be solved for $\bar{a}_0, \bar{a}_1, \bar{b}_1$; we have

$$1.4\bar{a}_0 + (-0.05)\bar{a}_1 + 0.02\bar{b}_1 = 1.4$$
$$(-0.05)\bar{a}_0 + 1.6\bar{a}_1 \quad + 0.24\bar{b}_1 = -1.3$$
$$0.02\bar{a}_0 + 0.24\bar{a}_1 \quad + 1.2\bar{b}_1 = \quad 0.16.$$

Dividing each equation by its *diagonal* coefficient we obtain the equivalent system

$$\bar{a}_0 - .036\bar{a}_1 + .014\bar{b}_1 = 1$$
$$-.031\bar{a}_0 + \quad \bar{a}_1 + .15\bar{b}_1 = -.81$$
$$.017\bar{a}_0 + \quad .2\bar{a}_0 + \quad \bar{b}_1 = .13.$$

We obtain, using $(I + M)^{-1} \approx I - M$,

$$\begin{pmatrix} \bar{a}_0 \\ \bar{a}_1 \\ \bar{b}_1 \end{pmatrix} \approx \begin{pmatrix} 1 & .036 & -.014 \\ .031 & 1 & -.15 \\ -.017 & -.2 & 1 \end{pmatrix} \begin{pmatrix} 1 \\ -.81 \\ .13 \end{pmatrix} = \begin{pmatrix} .97 \\ -.80 \\ .28 \end{pmatrix}$$

so that the sought after values of $\bar{a}_0, \bar{a}_1, \bar{b}_1$ are approximately

$$\bar{a}_0 = .97$$
$$\bar{a}_1 = -.80$$
$$\bar{b}_1 = .28.$$

Using these values, we can *plot* the resulting finite Fourier-series approximation

$$\bar{y}(t) = (.97)\bar{\varphi}_0(t) + (-.80)(.75) \sin (1.8)t + (.28)(.75) \cos (1.8)t$$
$$= .51 - .60 \sin 1.8t + .21 \cos 1.8t$$

to the given data for f. The result is shown sketched in Figure 5.15; the curve shown is $\bar{y}(t)$ and the circled points are the given data for f.

The data given in the table of values for f used in this example were *read off* the curve shown in Figure 5.14 after choosing a scale and were selected to represent the *main features* of the wave form. Compare the resulting three-term, Fourier-series approximation shown in Figure 5.15.

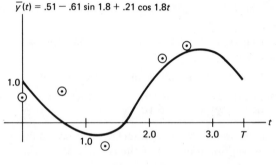

$$\bar{y}(t) = .51 - .61 \sin 1.8 + .21 \cos 1.8t$$

Figure 5.15

EXERCISE

Using a computer for the numerical work, find and *plot* a seven-term, Fourier-series approximation, in $S_3[T_n]$, to the curve shown in Figure 5.14 at the beginning of this section by choosing a scale and reading-off data points *approximately* chosen from the figure. Do this for a number of choices of T_n (always keeping $n > 7$). Study the resulting approximating curves to get some *feeling* for an *appropriate* choice of the set of data points to be used in the fitting.

apter 6

Discrete
approximation

6.1 Discrete Functions as Restrictions of Continuous Functions

A function f is *continuous* if it is *locally constant* at every point x in its domain; that is, if we can make $f(y)$ as *close* as we please to $f(x)$ simply by restricting y to be *sufficiently close* to x.

Let f be a continuous real-valued function with domain [0,1]. Thus f is locally constant at every x in [0,1]. We can approximate f by discrete functions of the form

$$\{[x_1, f(x_1)], [x_2, f(x_2)], \ldots, [x_n, f(x_n)]\}$$

for $0 \leq x_1 < x_2 < \cdots < x_n \leq 1$ by taking enough points $x_1, x_2, \ldots, x_n$ so that every x in [0,1] is sufficiently close to one of the points x_i ($i = 1, 2, \ldots, n$).

If T is a finite set of points in [0,1], then in an earlier section we have defined the restriction of f to T as the discrete function $f|_T$ whose domain is T and whose values on T are the same as those of f, so

$$f|_T(x_i) = f(x_i)$$

for every x_i in T. Thus $f|_T$ is a *discrete approximation* to the continuous function f.

121

6.2   Quadrature Methods

In this section we consider some numerical methods for approximating the value of the definite integral.

$$I(f) = \int_0^1 f(x)dx$$

for a continuous function f whose domain includes the *interval of integration* [0,1].

The more general appearing problem of finding

$$\int_a^b g(y)dy$$

can be put into the previous form by a change of the *variable of integration;* if we put

$$y = a + (b - a)x$$

then

$$x = \frac{y - a}{b - a}$$

and

$$\int_a^b g(y)dy = (b - a) \int_0^1 g[a + (b - a)x]dx$$

$$= (b - a) \int_0^1 f(x)dx$$

with

$$f(x) = g[a + (b - a)x].$$

The integral $I(f)$ is defined as the area under the graph of f between 0 and 1 (see Figure 6.1). The area of the shaded region is $I(f)$.

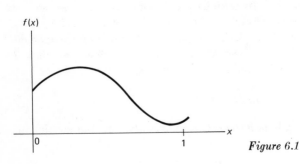

Figure 6.1

A continuous function f is locally constant, so we can approximate the area $I(f)$ by the sum of areas of the narrow rectangles with base lengths $1/n$ and heights $f(i/n)$ for $i = 1, 2, \ldots, n$, as shown in Figure 6.2. This results in the formula

$$S_n(f) = \sum_{i=1}^{n} f\left(\frac{i}{n}\right)\frac{1}{n}$$

for approximating $I(f)$. Notice that $S_n(f)$ is the *arithmetic mean* or *average value* of f on the discrete set $\{1/n, 2/n, \ldots, n/n\}$, namely

$$S_n(f) = \frac{1}{n}\sum_{i=1}^{n} f\left(\frac{i}{n}\right).$$

For each n, $S_n(f)$ is a number that lies between the minimum and maximum values of f on $[0,1]$. In fact, since f is continuous and so takes on all values between its minimum and maximum value, according to the mean-value theorem of calculus, there is some point $\bar{x}$ in $[0,1]$ (perhaps several) such that

$$I(f) = f(\bar{x}).$$

We will almost never know such a point, and it will usually be difficult to find one, even approximately. On the other hand, if f is a *linear* function, say

$$f(x) = a + bx$$

then

$$I(f) = f(\tfrac{1}{2}),$$

since, in this case,

$$I(f) = \int_0^1 (a + bx)dx = \left(ax + \frac{bx^2}{2}\right)\Big|_0^1 = a + b(\tfrac{1}{2}).$$

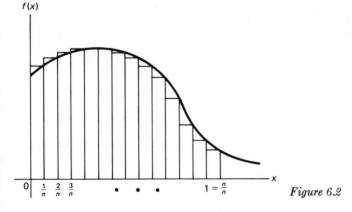

Figure 6.2

We could also get $I(f)$ for a linear function f by averaging the values of f at the end-points of the interval of integration. Thus

$$\int_0^1 (a + bx)dx = \tfrac{1}{2}\{(a + b\cdot 0) + (a + b\cdot 1)\}.$$

If the function f is *piecewise linear*, then the graph of f is a broken line (open polygon), say with connected line segments from 0 to x_1, x_1 to x_2, ..., x_k to 1, as shown in Figure 6.3. In such a case, $I(f)$ is the sum of areas of the trapezoids shown in Figure 6.3, namely

$$I(f) = x_1\left[\frac{f(0) + f(x_1)}{2}\right] + (x_2 - x_1)\left[\frac{f(x_1) + f(x_2)}{2}\right]$$
$$+ \cdots + (1 - x_k)\left[\frac{f(x_k) + f(1)}{2}\right].$$

Since a differentiable function is locally linear, we may use the formula above to good advantage in approximating $I(f)$ for differentiable f. To do this, we put

$$x_i = \frac{i}{n} \quad \text{for } i = 0, 1, 2, \ldots, n$$

and define the (summed version of the) *trapezoidal rule* by the formula

$$T_n(f) = \sum_{i=0}^n w_i f\left(\frac{i}{n}\right)$$

with

$$w_i = \begin{cases} \dfrac{1}{2n} & \text{if } i = 0 \quad \text{or} \quad i = n \\[2ex] \dfrac{1}{n} & \text{if } 0 < i < n. \end{cases}$$

See Figure 6.4.

The *weights* w_i in the formula $T_n(f)$ can be seen to arise from the formula for $I(f)$ in Figure 6.3. Notice that $\sum_{i=0}^n w_i = 1$ because $T_n(f)$ is exact for the

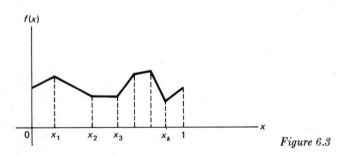

Figure 6.3

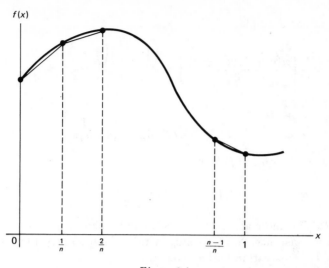

Figure 6.4

constant function $f(x) \equiv 1$. With practically no increase in computing time, we can use the formula $T_n(f)$ instead of $S_n(f)$ for approximating $I(f)$. When f is differentiable there is an improvement in accuracy, at least for big enough values of n. Compare Figures 6.2 and 6.4.

Consider, for instance, the example

$$I(f) = \int_0^1 x^2 dx$$

so that $f(x) = x^2$. Here f *is* differentiable and we can compare

$$S_n(f) = \frac{1}{n} \sum_{i=1}^n \left(\frac{i}{n}\right)^2$$

with

$$T_n(f) = \sum_{i=0}^n w_i \left(\frac{i}{n}\right)^2, \quad w_i = \begin{cases} \dfrac{1}{2n} & (i = 0, n) \\ \dfrac{1}{n} & (0 < i < n) \end{cases}$$

In this *example*, we have actually

$$S_n(f) = T_n(f) + \frac{1}{2n}.$$

It can be shown (by mathematical induction) that

$$\sum_{i=1}^n i^2 = \frac{n(n+1)(2n+1)}{6}$$

so that, in our example,

$$S_n(f) = \frac{1}{n^3}\left[\frac{n(n+1)(2n+1)}{6}\right] = \frac{\left(1+\frac{1}{n}\right)\left(2+\frac{1}{n}\right)}{6} = \frac{1}{3} + \frac{1}{2n} + \frac{1}{6n^2}$$

and so

$$T_n(f) = S_n(f) - \frac{1}{2n} = \frac{1}{3} + \frac{1}{6n^2}.$$

Now the exact result $I(f)$ is clearly

$$I(f) = \int_0^1 x^2 dx = \left.\frac{x^3}{3}\right|_0^1 = \frac{1}{3}$$

and so $T_n(f)$ is more accurate in this example than $S_n(f)$, for any n. When $n = 10$, for instance, $S_{10}(f)$ is only within about 15 percent of $I(f) = \frac{1}{3}$, whereas $T_{10}(f)$ is within half a percent.

EXERCISE

Use the trapezoidal rule $T_n(f)$ to compute an approximate value of

$$I = \int_0^1 e^{-x^2} dx.$$

Put $f(x) = e^{-x^2}$ and compute $T_n(f)$ for $n = 2, 4, 8, 16$. Compare results. Estimate the dependence of $I - T_n(f)$ on n. (*Hint:* Find the *least-squares* fit to the computed values of $T_n(f)$ of the form $T_n(f) = I + (C/n^2)$; put $\varphi_1(t) = 1$, $\varphi_2(t) = 1/t^2$ and use the data at $t = 2, 4, 8, 16$ to find the best-fitting values of I and C. Plot the resulting curve and the data.)

We can, of course, integrate polynomial functions (and many other *elementary functions*) exactly. If

$$p(x) = a_0 + a_1 x + \cdots + a_n x^n$$

then

$$I(p) = \int_0^1 p(x) dx = a_0 + \frac{a_1}{2} + \cdots + \frac{a_n}{n+1}.$$

We can also express $I(p)$ exactly in terms of a weighted sum of values of p at the points $0, 1/n, 2/n, \ldots, 1$. To do this we set

$$I(p) = \sum_{i=0}^{n} w_i p\left(\frac{i}{n}\right).$$

We can find *one set of weights* $w_0, w_1, w_2, \ldots, w_n$ which will work for *all* polynomials of degree n (or less). We can find a set of weights $w_0, w_1, w_2, \ldots, w_n$ such that

$$I(p) = \sum_{i=0}^{n} w_i p\left(\frac{i}{n}\right)$$

holds simultaneously for all the special polynomials of the form $p_k(x) = x^k$, $k = 0, 1, 2, \ldots, n$ (with $x^0 \equiv 1$) and then it will follow that, for any $a_0, a_1, \ldots, a_n,$

$$I(p) = \sum_{k=0}^{n} a_k I(p_k) = \sum_{k=0}^{n} a_k \sum_{i=0}^{n} w_i \left(\frac{i}{n}\right)^k = \sum_{i=0}^{n} w_i p\left(\frac{i}{n}\right).$$

We have a linear algebraic system to solve for the weights, namely

$$\sum_{i=0}^{n} w_i p_k\left(\frac{i}{n}\right) = I(p_k), \quad (k = 0,1, \ldots, n).$$

Now $I(p_k) = \int_0^1 x^k dx = 1/(k + 1)$ and $p_k(i/n) = (i/n)^k$ (with $p_0(x) \equiv 1$) and so the system of equations to be solved for $w_0, w_1, w_2, \ldots, w_n$ is

$$w_0 + w_1 + w_2 + \cdots + w_n = 1$$

$$\left(\frac{1}{n}\right)w_1 + \left(\frac{2}{n}\right)w_2 + \cdots + \left(\frac{n}{n}\right)w_n = \frac{1}{2}$$

$$\left(\frac{1}{n}\right)^2 w_1 + \left(\frac{2}{n}\right)^2 w_2 + \cdots + \left(\frac{n}{n}\right)^2 w_n = \frac{1}{3}$$

$$\vdots$$

$$\left(\frac{1}{n}\right)^n w_1 + \left(\frac{2}{n}\right)^n w_2 + \cdots + \left(\frac{n}{n}\right)^n w_n = \frac{1}{n + 1}.$$

For instance, when $n = 2$, we have

$$w_0 + w_1 + w_2 = 1$$
$$\tfrac{1}{2}w_1 + w_2 = \tfrac{1}{2}$$
$$\tfrac{1}{4}w_1 + w_2 = \tfrac{1}{3}$$

and we find that $w_0 = \tfrac{1}{6}, w_1 = \tfrac{4}{6}, w_2 = \tfrac{1}{6}$. (Of course, for a different value of n, we will get another, different set of values for $w_0, w_1, w_2, \ldots, w_n$.)

EXERCISE

Find the weights w_0, w_1, w_2, w_3 when $n = 3$. Check that the resulting formula gives the exact result for $I(f)$ when $f(x) = x^3$.

The *three-point* formula we have obtained (using $n = 2$) with the weights $w_0 = \tfrac{1}{6}, w_1 = \tfrac{4}{6}, w_2 = \tfrac{1}{6}$ is called *Simpson's rule:*

$$Sr(f) = \tfrac{1}{6}f(0) + \tfrac{4}{6}f(\tfrac{1}{2}) + \tfrac{1}{6}f(1).$$

When f is a polynomial of degree less than or equal to 2, we will have

$$Sr(f) = I(f);$$

that is, Simpson's rule will give exact results for the integral of second degree polynomials (*quadratic* functions of x).

For the example $f(x) = x^2$, we can check that

$$\begin{aligned}Sr(f) &= \tfrac{1}{6}\cdot(0)^2 + \tfrac{4}{6}(\tfrac{1}{2})^2 + \tfrac{1}{6}(1)^2 \\ &= 0 + \tfrac{1}{6} + \tfrac{1}{6} \\ &= \tfrac{1}{3} = \int_0^1 x^2 dx = I(f).\end{aligned}$$

EXERCISE

Show that Simpson's rule also gives *exact* results for *cubics* (polynomials of degree 3)!

We can make use of the identity

$$\int_a^b f(x)dx = (b-a)\int_0^1 f[a + (b-a)x]dx$$

to derive a *summed version of Simpson's rule* SR_N that will give exact results for the integral of any function which is quadratic (or cubic!) in x over each of the *subintervals* $[0,2h], [2h,4h], \ldots, [2(N-1)h,2Nh]$ with $h = 1/2N$.

We have

$$\int_{2ih}^{2ih+2h} f(x)dx = 2h\int_0^1 f(2ih + 2hx)dx$$

and we can apply $Sr(g_i)$ to $g_i(x) = f(2ih + 2hx)$ to get

$$\begin{aligned}Sr(g_i) &= \tfrac{1}{6}g_i(0) + \tfrac{4}{6}g_i(\tfrac{1}{2}) + \tfrac{1}{6}g_i(1) \\ &= \tfrac{1}{6}f(2ih) + \tfrac{4}{6}f[(2i+1)h] \\ &\quad + \tfrac{1}{6}f[(2i+2)h].\end{aligned}$$

Summing over $i = 0, 1, 2, \ldots, N-1$, we get

$$I(f) = \int_0^1 f(x)dx = \sum_{i=0}^{N-1}\int_{2ih}^{2ih+2h} f(x)dx$$

and we can approximate $I(f)$ by

$$SR_N(f) = \sum_{i=0}^{N-1} 2h[Sr(g_i)].$$

EXERCISE

Show that $SR_N(f)$ can be written

$$SR_N(f) = \frac{h}{3} \sum_{j=0}^{2N} W_j f(jh)$$

with

$$W_j = \begin{cases} 1 & \text{if } j = 0, 2N \\ 4 & \text{if } j \text{ is odd and } 0 < j < 2N \\ 2 & \text{if } j \text{ is even and } 0 < j < 2N. \end{cases}$$

Simpson's rule will give an approximation $SR_N(f)$, which is more accurate than $T_{2N}(f)$ given by the trapezoidal rule with the same number, $2N$, of function evaluations whenever the function f is *better* approximated locally (for that value of N) by quadratic (or cubic) functions than by a pair of linear functions.

EXERCISES

1. Use Simpson's rule to approximate

$$I = \int_0^1 e^{-x^2} dx.$$

Compute $SR_N(f)$, with $f(x) = e^{-x^2}$, for $N = 1, 2, 4$. Compare results with those obtained using the trapezoidal rule.

2. Write a subroutine to perform numerical integration of

$$\int_a^b f(x) dx$$

by the trapezoidal rule.

Try to make your program as general as possible; it should accept the following arguments:

1. Function $f(x)$ to integrate.
2. Lower limit of integration a.
3. Upper limit of integration b.
4. Number of *steps* N into which the interval of integration is to be divided.

It should return the numerical value of the approximated integral. Note that for complete generality you should code each function $f(x)$ to be integrated as a *separate* subprogram. Remember, your subroutine to approximate an integral is *general* only if it can be used, *without change*, for different functions, different limits, and different step sizes. Also try to write your program to make as few function evaluations as necessary.

3. Write a subroutine, just as above, for Simpson's rule.

4. Use your trapezoidal subroutine to find

$$\int_0^1 xe^{-x^2}dx \quad \text{and} \quad \int_0^1 x^2e^{-x^2}dx$$

for $N = 2, 4, 8, 16$.

5. Use your Simpson's subprogram to find the same integrals for $N = 1, 2, 4$.

6. Compare trapezoidal results to Simpson results. Also compare each set of approximations as N gets larger.

We consider next, *Gaussian quadrature*, which is very efficient for smooth functions that are *well* approximated locally by polynomials of high degree. Let us seek a quadrature formula of the form

$$\int_0^1 p(t)dt \approx \sum_{i=1}^N w_i p(t_i)$$

which is *exact* for *all* polynomials $p(t)$ of as high degree as possible for a fixed N. We have $2N$ parameters to play with in the quadrature formula, namely $w_1, w_2, \ldots, w_N$ and $t_1, t_2, \ldots, t_N$ and so we might expect to be able to choose them so that we will get exact results for all polynomials p of degree $2N - 1$ or less

$$p(t) = a_0 + a_1t + a_2t^2 + \cdots + a_{2N-1}t^{2N-1}$$

since there are $2N$ parameters (coefficients $a_0, a_1, \ldots, a_{2N-1}$) which may vary independently. We *can* do this, in fact, as follows.

We can find, first, a polynomial T_N of degree N which is orthogonal on $[0,1]$ to all polynomials of degree less than N. We seek

$$T_N(t) = b_0 + b_1t + \cdots + b_Nt^N$$

such that

$$\int_0^1 T_N(t)t^i dt = 0 \qquad (i = 0,1,2, \ldots, N - 1).$$

This gives a system of equations to be solved for $b_0, b_1, \ldots, b_N$:

$$b_0 + \tfrac{1}{2}b_1 + \cdots + \frac{1}{N + 1} b_N = 0$$

$$\tfrac{1}{2}b_0 + \tfrac{1}{3}b_1 + \cdots + \frac{1}{N + 2} b_N = 0$$

$$\vdots$$

$$\frac{1}{N} b_0 + \frac{1}{N + 1} b_1 + \cdots + \frac{1}{2N} b_N = 0.$$

There are two things *wrong* with this system of equations. First, it has infinitely many solutions; but that is not too bad because we only need one and if we set $b_N = 1$ we will have a nonsingular system

$$b_0 + \tfrac{1}{2}b_1 + \cdots + \frac{1}{N}\,b_{N-1} = -\frac{1}{N+1}$$

$$\tfrac{1}{2}b_0 + \tfrac{1}{3}b_1 + \cdots + \frac{1}{N+1}\,b_{N-1} = -\frac{1}{N+2}$$

$$\cdot$$
$$\cdot$$
$$\cdot$$

$$\frac{1}{N}\,b_0 + \frac{1}{N+1}\,b_1 + \cdots + \frac{1}{2N-1}\,b_{N-1} = -\frac{1}{2N}$$

which has a unique solution for $b_0, b_1, \ldots, b_{N-1}$ and will give, together with $b_N = 1$, suitable T_N.

The other thing that is *wrong* with the system — even after this change — is that it is *ill-conditioned* and a numerical solution, using Gaussian elimination, for instance, will need to carry a large number of decimal places (for even moderate values of N) to get good accuracy in the solution for $b_0, \ldots, b_{N-1}$. Another approach to finding T_N is to use the Gram-Schmidt process described earlier. The polynomial T_N has N distinct real zeros $t_1, t_2, \ldots, t_N$ in [0,1]. Now any polynomial $p(t)$ of degree $2N - 1$ or less can be written

$$p(t) = Q_{N-1}(t)T_N(t) + R_{N-1}(t)$$

where Q_{N-1} and R_{N-1} are polynomials of degree $N - 1$ or less (this can be verified by *long division* of $p(t)$ by $T_N(t)$).

Then we have

$$\int_0^1 p(t)dt = \int_0^1 Q_{N-1}(t)\,T_N(t)dt + \int_0^1 R_{N-1}(t)dt.$$

The integral

$$\int_0^1 Q_{N-1}(t)\,T_N(t)dt$$

is zero because T_N is *orthogonal* to *every* polynomial of degree $N - 1$ or less. Thus we have

$$\int_0^1 p(t)dt = \int_0^1 R_{N-1}(t)dt.$$

Now we want

$$\int_0^1 p(t)dt = \sum_{i=1}^{n} w_i p(t_i) = \int_0^1 R_{N-1}(t)dt$$

and

$$p(t_i) = Q_{N-1}(t_i)T_N(t_i) + R_{N-1}(t_i).$$

If we take $t_1, t_2, \ldots, t_N$ to *be the zeros of* T_N, then

$$p(t_i) = R_{N-1}(t_i) \qquad (i = 1,2, \ldots, N).$$

We then seek $w_1, w_2, \ldots, w_N$ such that (for $t_1, t_2, \ldots, t_N$ chosen as the zeros of T_N) the formula

$$\int_0^1 R_{N-1}(t)dt = \sum_{i=1}^N w_i R_{N-1}(t_i)$$

will be exact for *any* polynomial R_{N-1} of degree $N - 1$ or less. This will be the case if we choose $w_1, w_2, \ldots, w_N$ so that

$$\int_0^1 t^k dt = \sum_{i=1}^N w_i t_i^k \qquad (k = 0,1, \ldots, N - 1).$$

This gives the following linear algebraic system for $w_1, w_2, \ldots, w_N$:

$$w_1 + w_2 + \cdots + w_N = 1$$
$$t_1 w_1 + t_2 w_2 + \cdots + t_N w_N = \tfrac{1}{2}$$
$$\vdots$$
$$t_1^{N-1} w_1 + t_2^{N-1} w_2 + \cdots + t_N^{N-1} w_N = \frac{1}{N}.$$

EXERCISE

Find the weights w_1, w_2, and argument spacings t_1, t_2 for the two-point
(N = 2) Gaussian quadrature formula.

We can use a Gaussian quadrature formula for a given N to approximate an integral

$$\int_0^1 g(t)dt \approx \sum_{j=1}^N w_i g(t_i)$$

even when g is not a polynomial. The error will be the smallest difference between the integral of g and the integral of any polynomial of degree $2N - 1$ or less. If g is very well approximated by *some* polynomial of degree $2N - 1$ or less, then this error will be very small. We can also use the Gaussian quadrature method *in steps* (summed version). Putting

$$\int_0^1 g(t)dt = \sum_{j=1}^J \left[\frac{1}{J} \int_0^1 g\left(\frac{j-1}{J} + \frac{t}{J} \right)dt \right]$$

and applying the N-point Gaussian quadrature formula to each integral on the right-hand side, we obtain

$$\int_0^1 g(t)dt \approx \frac{1}{J} \sum_{j=1}^{J} \left[\sum_{i=1}^{N} w_i g\left(\frac{j-1}{J} + \frac{t_i}{J} \right) \right].$$

The best use of this formula for a given function g will involve choosing J and N so that g is well approximated by different polynomials of degree $2N - 1$ or less on *each* of the J intervals $[(j-1)/J, j/J]$ $(j = 1, 2, \ldots, J)$.

The Gaussian quadrature formula is sufficiently useful that weights $[w_i]$ and argument spacings $[t_i]$ have been tabulated for values of N to at least 16 (Lowans et al., 1943).

EXERCISE

Try the summed version of the two-point Gaussian quadrature formula you have found from the previous exercise on the integral

$$I = \int_0^1 e^{-x^2} dx$$

for $J = 1, 2, 4$. Compare results with those obtained from previous exercises using trapezoidal and Simpson's rules.

A method that can often be used for the efficient numerical approximation of definite integrals is the *integration of local finite Taylor series expansions*. For this, it is convenient to introduce the notation

$$(y)_k(x_i) = \frac{1}{k!} \frac{d^k y}{dx^k}\bigg|_{x=x_i} \qquad \text{where} \quad (y)_0(x_i) = y(x_i)$$

for the kth Taylor coefficient of y at x_i. Thus

$$y(x_i + t) = \sum_{k=0}^{K} (y)_k(x_i) t^k + R_K(y,x_i,t)$$

where

$$R_K(y,x_i,t) = (y)_{K+1}(\xi) t^{K+1}$$

for some ξ in $[x_i, x_i + t]$ whenever y is $K + 1$ times continuously differentiable. We have, in this case, for $x_i = ih = i/n$,

$$\int_0^1 y(x)dx = \sum_{i=0}^{n-1} \int_{x_i}^{x_i+h} y(x)dx$$

$$= \sum_{i=0}^{n-1} \int_0^h y(x_i + t)dt$$

$$= \sum_{i=0}^{n-1} \sum_{k=0}^{K} \frac{1}{k+1} (y)_k(x_i) h^{k+1} + \bar{R}_{K,n}(y)$$

where

$$\bar{R}_{K,n}(y) = \sum_{i=0}^{n-1} \int_0^h R_K(y, x_i, t) dt.$$

Notice that $\bar{R}_{K,n}(y)$ will be zero if y is a polynomial of degree K or less in each subinterval $[x_i, x_i + h]$.

As an example, consider the approximation of

$$I = \int_0^1 e^{-x^2} dx$$

by

$$TS_{n,K}(y) = h \sum_{i=0}^{n-1} \sum_{k=0}^{K} \frac{1}{k+1} (y)_k(x_i) h^k$$

with $y(x) = e^{-x^2}$.

The Taylor coefficients can be computed *recursively* as follows. We have

$$(y)_1(x) = e^{-x^2}(-2x) = -2xy(x)$$

$$(y)_2(x) = -\frac{2}{2}[x(y)_1(x) + (y)_0(x)]$$

$$\vdots$$

$$(y)_k(x) = -\frac{2}{k}[x(y)_{k-1}(x) + (y)_{k-2}(x)] \qquad \text{for } k = 2, 3, \ldots, K.$$

In order to compute one set of Taylor coefficients $(y)_0, (y)_1, \ldots, (y)_K$ for $x = x_i$, we require *one* evaluation of the integrand

$$y(x_i) = e^{-x_i^2}$$

and $2K$ multiplications plus $K - 1$ additions. If we evaluate the n polynomials in h

$$\sum_{k=0}^{K} \frac{1}{k+1} (y)_k(x_i) h^k \qquad (i = 0, 1, \ldots, n-1)$$

which occur in $TS_{n,K}(y)$ by the nesting method (perhaps *storing* the factors $1/(k+1)$, $k = 0, 1, \ldots, K$, in advance) we will require a total of no more than $2(n-1)K + 1$ multiplications and $(n-1)(K+1)$ additions and n evaluations of the integrand $y(x_i) = e^{-x_i^2}$, $i = 0, 1, \ldots, n-1$.

EXERCISES

1. Show that $n = 3$ approximately minimizes total operation count in the above example subject to requiring a fixed accuracy independent of K and n.

2. Try the method (carry out the computation using a computer, of course) on the integral

$$I = \int_0^1 e^{-x^2} dx$$

using all combinations of $K = 2, 4, 8, 12$ and $n = 2, 3, 4$. Compare results with those obtained previously using Simpson's rule, trapezoidal rule, and Gaussian quadrature.

It was not an accident that we were able to compute the Taylor coefficients of the function e^{-x^2} recursively. We can do this for any function that is composed of rational and elementary functions (sin, cos, log, exp, . . .).

Using the notation

$$(y)_k = \frac{1}{k!} \frac{d^k y}{dx^k}$$

as before, the following rules apply (Moore, 1966, Chapter 11):

$$(u + v)_k = (u)_k + (v)_k$$

$$(u - v)_k = (u)_k - (v)_k$$

$$(uv)_k = \sum_{j=0}^{k} (u)_j (v)_{k-j}$$

$$\left(\frac{u}{v}\right)_k = \left(\frac{1}{v}\right)\left[(u)_k - \sum_{j=1}^{k} (v)_j \left(\frac{u}{v}\right)_{k-j}\right]$$

$$(u^a)_k = \left(\frac{1}{u}\right) \sum_{j=0}^{k-1} \left(a - \frac{j(a+1)}{k}\right)(u)_{k-j}(u^a)_j$$

$$(e^u)_k = \sum_{j=0}^{k-1} \left(1 - \frac{j}{k}\right)(e^u)_j(u)_{k-j}$$

$$(\log_e u)_k = \left(\frac{1}{u}\right)\left[(u)_k - \sum_{j=1}^{k-1} \left(1 - \frac{j}{k}\right)(u)_j(\log_e u)_{k-j}\right]^*$$

$$(\sin u)_k = \sum_{j=0}^{k-1} \left(\frac{j+1}{k}\right)(\cos u)_{k-1-j}(u)_{j+1}$$

$$(\cos u)_k = -\sum_{j=0}^{k-1} \left(\frac{j+1}{k}\right)(\sin u)_{k-1-j}(u)_{j+1}$$

.
.
.

The sin and cos formulas should be used together as a pair even if only one is needed directly. We will derive two of these formulas. The rest of the deriva-

*The sum is deleted if $k = 1$.

tives will be left as an exercise for the reader. The formula for the kth Taylor coefficient of the product of two functions is essentially Leibnitz' formula

$$\frac{d^k(uv)}{dx^k} = \sum_{j=0}^{k} \binom{k}{j} \frac{d^j u}{dx^j} \frac{d^{k-j}v}{dx^{k-j}}$$

where the binomial coefficient

$$\binom{k}{j} = \frac{k!}{j!(k-j)!}$$

can be broken up into its three factors to obtain (dividing both sides of Leibnitz' formula by $k!$)

$$\frac{1}{k!}\frac{d^k(uv)}{dx^k} = \sum_{j=0}^{k} \left(\frac{1}{j!}\frac{d^j u}{dx^j}\right)\left(\frac{1}{(k-j)!}\frac{d^{k-j}v}{dx^{k-j}}\right)$$

which *is*, in our notation,

$$(uv)_k = \sum_{j=0}^{k} (u)_j (v)_{k-j}.$$

To derive the formula for $(e^u)_k$, we observe first that $(e^u)_1 = e^u(u)_1$; then we can apply the formula for the $(k-1)$st Taylor coefficient of a product to obtain

$$[(e^u)_1]_{k-1} = [e^u(u)_1]_{k-1} = \sum_{j=0}^{k-1} (e^u)_j[(u)_1]_{k-1-j}.$$

Now for any function v we have

$$[(v)_1]_{k-1} = k(v)_k$$

so we can simplify the expression above to obtain

$$(e^u)_k = \left(\frac{1}{k}\right)\sum_{j=0}^{k-1} (e^u)_j(k-j)(u)_{k-j}$$

$$= \sum_{j=0}^{k-1} \left(1 - \frac{j}{k}\right)(e^u)_j(u)_{k-j}$$

EXERCISE

Derive the rest of the recursion formulas given for kth Taylor coefficients.

We can illustrate the application of some of these formulas by considering the following example. Suppose we wish to compute a numerical approximation to

$$I = \int_0^1 e^{-\sin x} \log_e (1 + x)dx$$

using the method of the integration of local finite Taylor series expansions. Let $f(x) = e^{-\sin x} \log_e (1 + x)$; then we need an efficient means for computing the Taylor coefficients $(f)_k(x_i)$, $k = 0, 1, 2, \ldots, K$ to be used in the approximation

$$I \approx TS_{n,K}(f)$$

where

$$TS_{n,K}(f) = h \sum_{i=0}^{n-1} \sum_{k=0}^{K} \frac{1}{k+1} (f)_k(x_i) h^k$$

with $x_i = ih$ and $nh = 1$.

To do this, we introduce some auxiliary variables and form the expressions

$$T_1 = \sin x$$
$$T_2 = \log_e (1 + x)$$
$$T_3 = e^{-T_1}$$
$$T_4 = \cos x$$
$$f(x) = T_2 T_3.$$

Selecting the appropriate rule for each line in this list of expressions and making use of the fact that $(x)_0 = x$, $(x)_1 = 1$, $(x)_j = 0$ for $j > 1$, we obtain (for $k = 1, 2, \ldots, K$)

$$(T_1)_k = \frac{1}{k} (T_4)_{k-1}$$

$$(T_2)_k = \frac{1}{1+x} \left[(1 + x)_k - \left(1 - \frac{1}{k} \right) (T_2)_{k-1} \right]$$

$$(T_3)_k = \sum_{j=0}^{k-1} \left(1 - \frac{j}{k} \right) (T_3)_j (-T_1)_{k-j}$$

$$(T_4)_k = -\frac{1}{k} (T_1)_{k-1}$$

$$(f)_k(x) = \sum_{j=0}^{k} (T_2)_j (T_3)_{k-j}.$$

We may use these recursion relations to compute

$$(f)_k(x) \text{ at } x = x_i \quad \text{for } k = 0, 1, 2, \ldots, K.$$

Note that $(1 + x)_k = 0$ for $k > 1$ and $(1 + x)_1 = 1$.

We compute and store the following two-dimensional array (matrix) of quantities proceeding downward through the elements of a given column before continuing with the first element of the next column to the right.

$$
\begin{array}{ccccc}
T_1 & (T_1)_1 & (T_1)_2 & \cdot & (T_1)_K \\
T_2 & (T_2)_1 & (T_2)_2 & \cdot & (T_2)_K \\
T_3 & (T_3)_1 & (T_3)_2 & \cdot & (T_3)_K \\
T_4 & (T_4)_1 & (T_4)_2 & \cdot & (T_4)_K \\
f   & (f)_1   & (f)_2   & \cdot & (f)_K.
\end{array}
$$

The coefficients we seek are the the elements of the bottom row of the matrix. To obtain one set of Taylor coefficient values for $(f)_k, k = 0, 1, 2, \ldots, K$ at a particular $x = x_i$ in this way requires *one* evaluation of f (the first column of the computation) plus approximately $[3K(K + 1)]/2$ multiplications and additions.

6.3   Accuracy of Quadrature Methods

In Section 6.2 various quadrature methods are derived. Each of them has the property that it is exact for all polynomials up to a certain degree. A quadrature *method Q* of the form

$$
\int_0^1 f(x)dx \approx \sum_{i=1}^n w_i f(x_i)
$$

is *fixed* by the choice of $w_1, w_2, \ldots, w_n$ and $x_1, x_2, \ldots, x_n$.

Let us define the *error in Q applied to f* as

$$
E(Q,f) = \int_0^1 f(x)dx - \sum_{i=1}^n w_i f(x_i).
$$

Suppose Q is exact for all polynomials of degree k (or less). Then $E(Q,p_k) = 0$ for any polynomial p_k of degree k.

If f is a continuous $(k + 1)$st derivative, then $f(x)$ can be represented in the form

$$
f(x) = p_k(x) + r(x)
$$

where $p_k(x)$ is the polynomial of degree k given by the terms up to x^k in the Taylor series about $x = 0$ with remainder $r(x)$ for $f(x)$, where

$$
r(x) = \frac{1}{(k + 1)!} \frac{d^{k+1}f(\xi)}{dx^{k+1}} x^{k+1}
$$

for some ξ in the interval $[0,x]$.

Suppose that

$$
\max_{0 \le x \le 1} \left| \frac{1}{(k + 1)!} \frac{d^{k+1}f(x)}{dx^{k+1}} \right| \le C_{k+1},
$$

then $|r(x)| \leq C_{k+1}x^{k+1}$ for x in [0,1]. Now $E(Q,f)$, as defined, has the property that $E(Q,p_k + r) = E(Q,p_k) + E(Q,r)$. Since $E(Q,p_k) = 0$, we have

$$E(Q,f) = E(Q,r).$$

We can bound $|E(Q,r)|$ as follows.

If $k > 0$, then Q is exact for constant functions in particular, so

$$E(Q,1) = \int_0^1 1 \cdot dx - \sum_{i=1}^n w_i = 0$$

and it follows that

$$\sum_{i=1}^n w_i = 1.$$

Now

$$|E(Q,r)| \leq \left| \int_0^1 r(x)dx \right| + \left| \sum_{i=1}^n w_i r(x_i) \right|.$$

$$\leq \left| \int_0^1 r(x)dx \right| + \left(\max_{0 \leq x \leq 1} |r(x)| \right) \sum_{i=1}^n w_i$$

$$\leq C_{k+1}\left(\frac{1}{k+2} \right) + C_{k+1}.$$

Thus we have the bound

$$|E(Q,f)| \leq \left(1 + \frac{1}{k+2} \right)C_{k+1}.$$

To get a bound for the error in the *summed version* of a quadrature method Q we proceed as follows. We can write the integral to be approximated as

$$\int_0^1 f(x)dx = \sum_{j=1}^J \int_{(j-1)/J}^{j/J} f(x)dx$$

$$= \sum_{j=1}^J \left[\frac{1}{J} \int_0^1 f_j(x)dx \right]$$

where $f_j(x) = f[(j-1)/(J) + x/J]$.

If we apply the method Q to each of the functions f_j $(j = 1,2,\ldots,J)$, we obtain the *summed version of Q*

$$\int_0^1 f(x)dx \approx \sum_{j=1}^J \left[\frac{1}{J} \sum_{i=1}^n w_i f_j(x_i) \right].$$

Define the *J-step error in Q applied to f* as

$$E_J(Q,f) = \int_0^1 f(x)dx - \sum_{j=1}^J \left[\frac{1}{J} \sum_{i=1}^n w_i f_j(x_i) \right].$$

Thus the previous $E(Q,f)$ is the same as the one-step error $E_1(Q,f)$. Clearly we have

$$E_J(Q,f) = \sum_{j=1}^{J} \left[\frac{1}{J}\, E(Q,f_j) \right].$$

Now we have, for each of the functions f_j $(j = 1,2,\ldots,J)$,

$$\frac{df_j(x)}{dx} = \frac{df\left(\dfrac{j-1}{J} + \dfrac{x}{J}\right)}{dx}\left(\frac{1}{J}\right)$$

$$\frac{d^2f_j(x)}{dx^2} = \frac{d^2f\left(\dfrac{j-1}{J} + \dfrac{x}{J}\right)}{dx^2}\left(\frac{1}{J}\right)^2$$

$$\vdots$$

$$\frac{d^{k+1}f_j(x)}{dx^{k+1}} = \frac{d^{k+1}f\left(\dfrac{j-1}{J} + \dfrac{x}{J}\right)}{dx^{k+1}}\left(\frac{1}{J}\right)^{k+1}.$$

It follows that for each $j = 1, 2, \ldots, J$ we have

$$\frac{1}{(k+1)!}\max_{0\le x\le 1}\left|\frac{d^{k+1}f_j(x)}{dx^{k+1}}\right| \le C_{k+1}\left(\frac{1}{J}\right)^{k+1}$$

and so

$$|E(Q,f_j)| \le \left(1 + \frac{1}{k+2}\right)C_{k+1}\left(\frac{1}{J}\right)^{k+1}.$$

Now

$$E_J(Q,f) = \sum_{j=1}^{J}\left[\frac{1}{J}\, E(Q,f_j)\right],$$

so we have derived the general bound

$$|E_J(Q,f)| \le \left(1 + \frac{1}{k+2}\right)C_{k+1}\left(\frac{1}{J}\right)^{k+1}.$$

The bound was derived using *only* the information that Q is exact for polynomials of degree k. If we know more about a particular Q, we can often find *sharper* bounds.

For the summed version of the trapezoidal rule, T, we can put $k = 1$ in the above general error bound and obtain

$$|E_J(T,f)| \le \frac{2}{3}\max_{0\le x\le 1}\left|\frac{d^2f(x)}{dx^2}\right|\left(\frac{1}{J}\right)^2.$$

We can compare this bound with the actual error for the example $f(x) = x^2$ (see Section 6.2). We have, for this example,

$$T_J(f) = \frac{1}{3} + \frac{1}{6}\left(\frac{1}{J}\right)^2$$

and

$$\int_0^1 x^2 dx = \tfrac{1}{3}$$

so that the actual error is $\frac{1}{6}(1/J)^2$. The general bound gives (since we may take $C_2 = 1$ in this example)

$$|E_J(T,f)| \leq \frac{4}{3}\cdot\left(\frac{1}{J}\right)^2.$$

For Simpson's rule, which is exact for polynomials of degree 3, the general bound gives

$$|E_J(S,f)| \leq \frac{1}{20} \max_{0 \leq x \leq 1} \left|\frac{d^4 f(x)}{dx^4}\right|\left(\frac{1}{J}\right)^4.$$

EXERCISE

Apply the general bound to Simpson's rule for the example $\int_0^1 e^{-x^2} dx$. Find a value of J for which the error will be less than 0.001.

If we *assume* that an estimate of the error $E_J(S,f)$ is given by

$$E_J(S,f) = C\left(\frac{1}{J}\right)^4$$

for *some* constant C, then we can estimate C computationally as follows. We compute S_J for two different values of J, say J_1 and J_2. We then have

$$I - S_{J_1} = C\left(\frac{1}{J_1}\right)^4$$

$$I - S_{J_2} = C\left(\frac{1}{J_2}\right)^4$$

where I is the unknown exact value of

$$\int_0^1 f(x)dx$$

and S_{J_1} and S_{J_2} are, respectively, the J_1 and J_2-step Simpson's rule results. We can eliminate the unknown I by subtracting these two equations to obtain

$$S_{J_2} - S_{J_1} = C\left[\left(\frac{1}{J_1}\right)^4 - \left(\frac{1}{J_2}\right)^4\right].$$

If we now solve this equation for C, we can estimate from

$$E_J(S,f) = C\left(\frac{1}{J}\right)^4$$

the value of J; we would then need to make $E_J(S,f)$ less than a given quantity.

EXERCISE

Suppose that by dividing $[0,1]$ into J equal parts, Simpson's rule (summed version) gives approximate values

J	S_J
4	3.221
8	3.002

to an integral $I = \int_0^1 f(x)dx$. For what value of J will S_J be accurate to four decimal places?

The N-*point Gaussian* quadrature method (see Section 6.2) is exact for polynomials of degree $2N - 1$ and the general bound gives, putting $k = 2N - 1$,

$$|E_J(\text{N-point } G., f)| \le \left(1 + \frac{1}{2N+1}\right)C_{2N}\left(\frac{1}{J}\right)^{2N}$$

where

$$C_{2N} = \max_{0 \le x \le 1}\left|\frac{1}{(2N)!}\frac{d^{2N}f(x)}{dx^{2N}}\right|.$$

EXERCISE

Assume that the error in N-point Gaussian quadrature is of the form

$$E_J(\text{N-point } G., f) = C\left(\frac{1}{J}\right)^{2N}$$

for some constant C. Show how to estimate C computationally by computing with two different values of J. Using $J = 2$ and $J = 4$ with $N = 2$ for $f(x) = e^{-x^2}$, estimate the error in E_2 (two-point $G., e^{-x^2}$).

Another way to estimate the error in approximating integrals is to use *interval computation*. We can compute an upper bound to the error in approximating

$$I(y) = \int_0^1 y(x)dx$$

by the local Taylor series approximation $TS_{n,K}(y)$ (discussed in Section 6.2), for instance, as follows. We have

$$I(y) - TS_{n,K}(y) = \bar{R}_{K,n}(y)$$

where

$$\bar{R}_{K,n}(y) = \sum_{i=0}^{n-1} \int_0^h R_K(y,x_i,t)dt$$

and

$$R_K(y,x_i,t) = (y)_{K+1}(\xi)t^{K+1}$$

for some ξ in $[x_i, x_i + t]$. Now we can evaluate $(y)_{K+1}$ using recursion formulas as explained in Section 6.2 not only in real arithmetic but in interval arithmetic as well (with appropriate *interval* subroutines for the elementary functions when needed; see Moore, 1966). If we do this for $(y)_{K+1}([x_i, x_i + h])$, then we will have

$$(y)_{K+1}(\xi) \in (y)_{K+1}([x_i, x_i + h])$$

whenever $\xi \in [x_i, x_i + t]$ and $t \in [0,h]$. Thus we obtain

$$R_K(y,x_i,t) \in (y)_{K+1}([x_i, x_i + h])t^{K+1}$$

and

$$\int_0^h R_K(y,x_i,t)dt \in (y)_{K+1}([x_i, x_i + h])\frac{h^{K+2}}{K+2}$$

and so the remainder satisfies

$$\bar{R}_{K,n}(y) \in \frac{h^{K+2}}{K+2} \sum_{i=0}^{n-1} (y)_{K+1}([x_i, x_i + h]).$$

For example, if $y(x) = e^{-x^2}$, we have (from Section 6.2) the recursion formula

$$(y)_k(x) = -\frac{2}{k}(x(y)_{k-1}(x) + (y)_{k-2}(x))$$

for $k = 1, 2, \ldots$ from which we can compute in interval arithmetic the quantities

$$(y)_{K+1}([x_i, x_i + h]) \qquad (i = 0, 1, \ldots, n - 1)$$

for a given choice of K. Even using the crude bound

$$y(x) = e^{-x^2} \in [0,1] \quad \text{for } x \in [0,1]$$

we can obtain a good estimate (and a strict upper bound) for the remainder as follows. Put $x_i = ih$ for $i = 0, 1, \ldots, n - 1$; $nh = 1$.

We have $(y)_0([ih,(i + 1)h]) \in [0,1]$. For ease of notation, define

$$Y_{k,i} = (y)_k([ih,(i + 1)h]);$$

then

$$Y_{0,i} \in [0,1] \qquad (i = 0,1, \ldots, n - 1)$$

and

$$Y_{1,i} \in -2[ih,(i + 1)h][0,1]$$

and so

$$Y_{1,i} \in [-2(i + 1),0]h$$

and, for $k > 1$,

$$Y_{k,i} \in \frac{-2}{k} ([ih,(i + 1)h]Y_{k-1,i} + Y_{k-2,i}).$$

From these relations, we can determine that for all $i = 0, 1, 2, \ldots, n - 1$, we have

$$Y_{1,i} \in [-2,0]$$
$$Y_{2,i} \in [-1,2]$$
$$Y_{3,i} \in [-\tfrac{4}{3},2]$$
$$Y_{4,i} \in [-2,\tfrac{7}{6}]$$

.

.

.

If the maximum width of the interval $Y_{k,i}$ for any $i = 0, 1, \ldots, n - 1$ is denoted by w_k, then it can be shown that, in this example,

$$w_k \leq \frac{4}{k} w_{k-1} \qquad (k > 1).$$

From this it follows that the w_k decrease with increasing k for $k > 4$. In fact, we can show that for all k and all i we have

$$Y_{k,i} \in [-2,2].$$

Therefore we have the following strict error bound:

$$\bar{R}_{K,n}(y) \in \frac{h^{K+1}}{K + 2} [-4,4]$$

or, for this example,

$$|I(y) - TS_{n,K}(y)| \leq \frac{4}{K + 2}\left(\frac{1}{n}\right)^{K+1}.$$

A more precise interval computation would yield a slightly sharper bound.

Recall, from Section 6.2, that the computation of $TS_{n,K}(y)$ for this example (where $y = e^{-x^2}$) was shown to require about $n(2K + E)$ multiplications, where E is the number of multiplications, required to evaluate the integrand, e^{-x^2}, once.

We can find an *optimal choice of* the *parameters* K and n by requiring that $n(2K + E)$ be minimum for all K and n which satisfy

$$\frac{4}{K + 2}\left(\frac{1}{n}\right)^{K+1} = \epsilon;$$

in other words, we seek K and n such that for a given *error tolerance* ϵ we minimize the estimated number of multiplications.

We can find the optimal n and K as follows. We first treat n and K as continuous variables; then

$$n(K) = \left(\frac{4}{\epsilon(K + 2)}\right)^{1/(K+1)}$$

and we seek the minimum value of

$$M(K) = (2K + E)\left(\frac{4}{\epsilon(K + 2)}\right)^{1/(K+1)}.$$

If we put $\epsilon = 10^{-6}$, for instance, then

$$M(2) = (4 + E)10^2 = 400 + 100E$$

$$M(5) = (10 + E)\left(\frac{4}{7}\right)^{1/6} 10 \approx 100 + 10E$$

$$M(11) = (22 + E)\left(\frac{4}{13}\right)^{1/12} 10^{1/2} \approx 70 + 3E$$

$$M(17) = (34 + E)\left(\frac{4}{19}\right)^{1/18} 10^{1/3} \approx 73 + 2E.$$

$$\vdots$$

The coefficient of E here is the value of $n(K)$ needed for $\epsilon = 10^{-6}$. We cannot get $n(K)$ down to 1 without increasing the constant term, $2K$, to enormous size because we then require that

$$\frac{4}{K + 2}\left(\frac{1}{1}\right)^{K+1} = 10^{-6}$$

so we would have, for $n = 1$,

$$K = 4 \cdot 10^6 - 2.$$

We conclude that the *optimal* choice of n and K when $\epsilon = 10^{-6}$ in this example is near $n = 2$, $K = 17$.

6.4   The Initial Value Problem for Ordinary Differential Equations

Discretization plays a role as the basis for many numerical methods for differential equations and integral equations (see Henrici, 1962; Babuska, 1966; and Forsythe and Wasow, 1960). If we approximate derivatives by finite-difference quotients or integrals by finite sums, then we can replace differential or integral equations by approximating *difference equations*. For instance, the differential equation

$$\frac{dN(t)}{dt} = aN(t)$$

can be approximated by the difference equation

$$\frac{N(t + \Delta t) - N(t)}{\Delta t} = aN(t)$$

in which we have approximated the derivative $dN(t)/dt$ by the *difference quotient* shown.

We could interpret the difference equation as describing (for $a > 0$) successive increases in the value of $N(t)$ at the discrete *times* Δt, $2\Delta t$, $3\Delta t$, ... , giving a $100a$ percent increase in $N(t)$ per unit time *compounded every* Δt. We have

$$N(t + \Delta t) = (1 + a\Delta t)N(t)$$

so that

$$N(k\Delta t) = (1 + a\Delta t)^k N(0).$$

If we arrange that $k\Delta t = 1$, then we are compounding k times per unit time and

$$N(1) = (1 + a\Delta t)^{1/\Delta t} N(0).$$

The differential equation correspondingly describes a continuous increase in the value of $N(t)$ at the *percentage rate* $100a$ per unit time *compounded* continuously. We have

$$\frac{dN(t)}{N(t)} = a \, dt$$

so that

$$\int_{N(0)}^{N(1)} \frac{dN(t)}{N(t)} = \int_0^1 a \, dt$$

or

$$\log_e N(t)\Big|_{N(0)}^{N(1)} = a$$

or

$$\log_e N(1) - \log_e N(0) = \log_e \left(\frac{N(1)}{N(0)}\right) = a$$

and so

$$N(1) = e^a N(0).$$

The quantity

$$(1 + a\Delta t)^{1/\Delta t}$$

is less than e^a but approaches e^a in the limit as $\Delta t \to 0$. When $a = 1$, for instance, we have

$$(1 + a\Delta t)^{1/\Delta t} = \begin{cases} 2 & \text{for } \Delta t = 1 \\ 2.25 & \text{for } \Delta t = \frac{1}{2} \\ 2.59\ldots & \text{for } \Delta t = 0.1 \end{cases}$$

whereas

$$e^a = e^1 = \lim_{\Delta t \to 0} (1 + \Delta t)^{1/\Delta t} = 2.7\ldots.$$

The same differential equation and associated difference equation just discussed can also be used to describe, approximately, population growth. If we have at time t a count of population $N(t)$ and if during the next interval of time of duration Δt there are B births, then the *birth rate* b could be *defined* as

$$b = \frac{B}{\Delta t N(t)}.$$

If, during the same interval, there are D deaths in the population, we could similarly define the *death rate* c as

$$c = \frac{D}{\Delta t N(t)}.$$

The total change in population during the interval is then

$$N(t + \Delta t) - N(t) = B - D = (b - c)\Delta t N(t)$$

or

$$\frac{N(t + \Delta t) - N(t)}{\Delta t} = aN(t)$$

where $a = b - c$.

In this case, we could regard the *differential equation*

$$\frac{dN(t)}{dt} = aN(t)$$

as an approximation to the difference equation! The *difference* equation has the solution

$$N(k\Delta t) = (1 + a\Delta t)^k N(0)$$

expressing the population at time $k\Delta t$ in terms of the population, $N(0)$, at *time zero* (whenever we begin counting, say). The corresponding differential equation has the solution

$$N(t) = e^{at}N(0)$$

as can be verified by differentiation

$$\frac{dN(t)}{dt} = e^{at}aN(0) = aN(t).$$

The solution to the differential equation at time $t = k\Delta t$ is

$$e^{at}N(0) = e^{ak\Delta t}N(0).$$

Since $(1 + a\Delta t)^{1/\Delta t}$ is approximated by e^a for small Δt, then

$$(1 + a\Delta t)^k = [(1 + a\Delta t)^{1/\Delta t}]^{k\Delta t}$$

is approximated by $(e^a)^{k\Delta t} = e^{ak\Delta t}$ for small Δt. Thus the difference equation and the differential equation have comparable solutions which, at a given value of t, come close to each other for small Δt.

EXERCISE

Discuss the behavior of $N(t)$ for large t for the three cases: $B > D$, $B = D$, $B < D$.

The differential equation

$$\frac{dN(t)}{dt} = aN(t)$$

also can be written as an *integral equation* obtained by integrating both sides of the differential equation from $t = 0$ to $t = T$

$$N(T) = N(0) + \int_0^T aN(t)dt.$$

We know the integral has the property that

$$\int_0^{T+\Delta t} aN(t)dt = \int_0^T aN(t)dt + \int_T^{T+\Delta t} aN(t)dt;$$

therefore the integral equation implies that

$$N(T + \Delta t) = N(0) + \int_0^{T+\Delta t} aN(t)dt = N(T) + \int_T^{T+\Delta t} aN(t)dt.$$

Using the trapezoidal approximation to the integral,

$$\int_T^{T+\Delta t} aN(t)dt \approx \Delta t\left[\frac{aN(T) + aN(T + \Delta t)}{2}\right],$$

and we obtain the following difference equation discretization of the integral equation

$$N(T + \Delta t) = N(T) + \Delta t\left[\frac{aN(T) + aN(T + \Delta t)}{2}\right].$$

Notice that $N(T + \Delta t)$ occurs on both sides of the difference equation. Since it occurs linearly here, we can solve explicitly for $N(T + \Delta t)$ to obtain the *explicit* difference equation

$$N(T + \Delta t) = \left(1 - \frac{a\Delta t}{2}\right)^{-1}\left(1 + \frac{a\Delta t}{2}\right)N(T).$$

Using this difference equation, we obtain

$$N(k\Delta t) = \left(\frac{1 + \dfrac{a\Delta t}{2}}{1 - (a\Delta t/2)}\right)^k N(0).$$

The exact solution of the integral equation at $t = k\Delta t$ is $e^{k\Delta t}N(0)$ and so it is interesting to compare the approximation to $e^{k\Delta t}$ given by

$$\left(\frac{1 + \dfrac{a\Delta t}{2}}{1 - (a\Delta t/2)}\right)^k$$

with that given by $(1 + a\Delta t)^k$ (resulting from the previously considered difference equation: $N(t + \Delta t) = (1 + a\Delta t)N(t)$).

We note that

$$\left(\frac{1 + \dfrac{a\Delta t}{2}}{1 - (a\Delta t/2)}\right)^k = \left[\left(\frac{1 + \dfrac{a\Delta t}{2}}{1 - (a\Delta t/2)}\right)^{1/\Delta t}\right]^{k\Delta t}$$

so that the comparison comes down to examining the approximations to $e = 2.718\ldots$ given by $(1 + a\Delta t)^{1/\Delta t}$ and by

$$\left(\frac{1 + \dfrac{a\Delta t}{2}}{1 - (a\Delta t/2)}\right)^{1/\Delta t}.$$

When $a = 1$, for instance, we have the following comparison:

Δt	$(1 + \Delta t)^{1/\Delta t}$	$\left(\dfrac{1 + \dfrac{\Delta t}{2}}{1 - \Delta t/2} \right)^{1/\Delta t}$
1	2	3
0.5	2.25	2.77...
0.1	2.59...	2.716...

The second difference equation gives more accurate approximation than the first for reasonably small Δt (say $a\Delta t < 1$) in this numerical example. *Systems* of differential equations of the form

$$\frac{dN_i}{dt} = e_i N_i + \sum_{s=1}^{n} a_{is} N_i N_s \qquad (i = 1, 2, \ldots, n)$$

have been used in mathematical analyses of ecological, biological, and chemical processes in which $N_i(t)$ represents the population (or density or concentration) of the ith species (or compound) at time t (for instance, in a system of interacting biological predator-prey relationships, *food-web* relationships, or interdependent chemical concentrations and reaction rates); see, for example, American Mathematical Society, 1968.

Such a system is included as a special case of the following even more general system:

$$\dot{x}_1 = \frac{dx_1(t)}{dt} = f_1(t, x_1, x_2, \ldots, x_n)$$

$$\dot{x}_2 = \frac{dx_2(t)}{dt} = f_2(t, x_1, x_2, \ldots, x_n)$$

$$\vdots$$

$$\dot{x}_n = \frac{dx_n(t)}{dt} = f_n(t, x_1, x_2, \ldots, x_n).$$

We can write such a system in briefer form as a *vector differential equation*

$$\frac{dx(t)}{dt} = f(t, x)$$

or

$$\dot{x} = f(t, x)$$

using the vector notation

$$x = [x_1(t), x_2(t), \ldots, x_n(t)]$$

and

$$f = (f_1, f_2, \ldots, f_n).$$

In an *initial value problem* we are given a (vector) differential equation

$$\dot{x} = f(t,x)$$

and initial conditions (vector x_0)

$$x = x_0 \quad \text{at } t = t_0$$

and we are to find a value or values of *the* solution $x(t)$ (if it exists, uniquely) for $t > t_0$. We can write this problem in mathematically equivalent form as the (vector) integral equation

$$x(t) = x_0 + \int_{t_0}^{t} f[t',x(t')]dt',$$

which represents the *system* of integral equations

$$x_1(t) = x_1(t_0) + \int_{t_0}^{t} f_1[t',x_1(t'), \ldots, x_n(t')]dt'$$

$$x_2(t) = x_2(t_0) + \int_{t_0}^{t} f_2[t',x_1(t'), \ldots, x_n(t')]dt'$$

.

.

.

$$x_n(t) = x_n(t_0) + \int_{t_0}^{t} f_n[t',x_1(t'), \ldots, x_n(t')]dt'.$$

If f_i is continuous for each $i = 1, 2, \ldots, n$, then a continuous solution $x(t)$ of the integral equation (each $x_i(t)$ continuous, $i = 1, 2, \ldots, n$) is also a solution of the initial value problem. This follows directly from the fundamental theorem of calculus. Since, for $t_0 < t_1 < \cdots < t_k < t_{k+1}$, we have

$$x(t_{k+1}) = x_0 + \int_{t_0}^{t_{k+1}} f[t',x(t')]dt'$$

$$= x_0 + \int_{t_0}^{t_k} f[t',x(t')]dt' + \int_{t_k}^{t_{k+1}} f[t',x(t')]dt'$$

$$= x(t_k) + \int_{t_k}^{t_{k+1}} f[t',x(t')]dt',$$

it follows that we can solve an initial value problem in a step-by-step fashion. We could approximate the integral

$$\int_{t_k}^{t_{k+1}} f[t',x(t')]dt'$$

by

$$(t_{k+1} - t_k)f[t_k, x(t_k)],$$

to obtain the approximating difference equation

$$x(t_{k+1}) = x(t_k) + (t_{k+1} - t_k)f[t_k, x(t_k)].$$

The use of this difference equation to compute approximate solutions to the initial value problem is referred to as *Euler's method.* If we denote an approximate solution by $X(t)$, then we can write one step of Euler's method in the simpler form

$$X(t + \Delta t) = X(t) + \Delta t f[t, X(t)].$$

Consider the nonlinear differential equation

$$\frac{dx(t)}{dt} = x^2(t)$$

with the initial condition $x(0) = 1$. For this example, Euler's method gives

$$X(t + \Delta t) = X(t) + \Delta t X^2(t).$$

Let T be fixed for the moment with $T > 0$ and consider Δt and k such that $k\Delta t = T$. We can use Euler's method to compute successively, beginning with $X(0) = 1$, the numbers

$$X(\Delta t), \quad X(2\Delta t), \ldots, X(k\Delta t)$$

in order to obtain $X_k = X(k\Delta t)$ as an approximation to the solution of the differential equation at $t = T$; and then we can investigate what happens to X_k as k gets larger and Δt gets smaller, keeping $k\Delta t = T$.

For simplicity we will use the notation $X_j = X(j\Delta t)$ for $j = 0, 1, 2, \ldots, k$. Suppose we take $T = 2$ and we compute

$$X_1, X_2, \ldots, X_k$$

using $\Delta t = T/k = 2/k$. In this special case, Euler's method can be written

$$X_{j+1} = X_j + \frac{2}{k} X_j^2 \qquad (j = 0, 1, 2, \ldots, k - 1)$$

with $X_0 = 1$.

Let us compute X_k for several choices of k and see what happens.

For $k = 1$ we have ($\Delta t = 2$)

$$X_k = X_1 = X_0 + \frac{2}{1} X_0^2 = 3.$$

For $k = 2$ we have ($\Delta t = 1$)

$$X_1 = X_0 + \frac{2}{2} X_0^2 = 2$$

$$X_k = X_2 = X_1 + \frac{2}{2} X_1^2 = 6.$$

For $k = 4$ we have ($\Delta t = 0.5$)

$$X_1 = 1 + 0.5(1)^2 = 1.5$$
$$X_2 = 1.5 + 0.5(1.5)^2 = 2.625$$
$$X_3 = X_2 + 0.5\,X_2{}^2 = 6.07\ldots$$
$$X_k = X_4 = X_3 + 0.5\,X_3{}^2 = 24.4\ldots.$$

It does not appear that X_k is converging to anything as k increases (as Δt decreases). In fact, $X_k \to \infty$ as $k \to \infty$ in this example for $T = 2$. Suppose we consider values of T other than 2 and use $\Delta t = T/k$; then we have

$$X_{j+1} = X_j + \frac{T}{k} X_j{}^2 \qquad (j = 0, 1, 2, \ldots, k - 1)$$

with $x_0 = 1$. For instance, for $T = 0.5$, we have: for $k = 1$ ($\Delta t = 0.5$)

$$X_k = X_1 = 1 + \frac{0.5}{1} \cdot 1^2 = 1.5;$$

for $k = 2$ ($\Delta t = 0.25$)

$$X_1 = 1 + 0.25(1)^2 = 1.25$$
$$X_k = x_2 = 1.25 + 0.25(1.25)^2 = 1.64\ldots$$

for $k = 5$ ($\Delta t = 0.1$)

$$X_1 = 1 + 0.1(1)^2 = 1.1$$
$$X_2 = 1.1 + 0.1(1.1)^2 = 1.221$$
$$X_3 = 1.221 + 0.1(1.221)^2 = 1.370\ldots$$
$$X_4 = 1.370\ldots + 0.1(1.370\ldots)^2 = 1.55\ldots$$
$$X_5 = X_5 = 1.55\ldots + 0.1(1.55\ldots)^2 = 1.79\ldots.$$

This time X_k is growing more slowly with increasing k, and it turns out that for $T = 0.5$, $\lim\limits_{k \to \infty} X_k = 2$.

EXERCISE

Recompute (on a computer) X_k, for $k = 10, 20, 40$, with $\Delta t = 0.5/k$; use $X_{j+1} = X_j + \Delta t X_j{}^2 (j = 0,1,2,\ldots, k - 1)$ with $X_0 = 1$ as before. Estimate the error $E_k = 2 - X_k$ as a function of Δt.

What is happening is the following. The differential equation

$$\frac{dx(t)}{dt} = x^2(t)$$

with the initial condition $x(0) = 1$ has the exact solution

$$x(t) = \frac{1}{1 - t}.$$

The difference equation

$$X_{j+1} = X_j + \frac{T}{k} X_j^2 \qquad (j = 0,1,\ldots,k-1)$$

with $X_0 = 1$ will give an X_k less than $1/(1-T)$ provided $T < 1$. For $T < 1$, it can be shown that $\lim_{k \to \infty} X_k = 1/(1-T)$; however, for $T > 1$, $\lim_{k \to \infty} X_k = \infty$.

In order to obtain conditions that will guarantee the convergence of discrete approximations to solutions of differential equations, we must somehow obtain bounds on the range of values of the solution over a domain of interest (if the solution *is* bounded over the chosen domain).

For the example under discussion this can be done as follows. Consider the range of values of the derivative $dx(t)/dt = x^2(t)$ when x is allowed to range over the interval $[1,b]$, $(b > 1)$. If $x(t)$ stays in $[1,b]$, then $dx(t)/dt$ stays in the interval $[1,b^2]$.

Now consider Figure 6.5. If we draw a pair of straight lines from $t = 0$, $x = 1$ with slopes 1 and b^2, then *to the point $\bar{t}$* where one of these lines gets out of the strip between $x = 1$ and $x = b$, the differential equation will have a solution lying within the wedge-shaped region; because we can write the differential equation as an integral equation

$$x(t) = 1 + \int_0^t x^2(t')dt'$$

and if $x(t')$ lies in $[1,b]$ for t' in $[0,\bar{t}]$, then $x^2(t')$ lies in $[1,b^2]$ and

$$\int_0^t x^2(t')dt'$$

lies in the interval $[1,b^2]t$ and so we can conclude that $x(t)$ lies in the wedge-shaped region described by

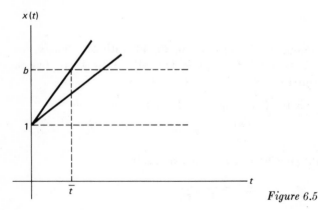

Figure 6.5

$$1 + [1,b^2]t \qquad \text{for } 0 \le t \le \overline{t}.$$

That is, for $0 \le t \le \overline{t}$, we have $1 + t \le x(t) \le 1 + b^2 t$.

In order to find a relation between b and $\overline{t}$ we must satisfy the condition

$$1 + b^2 \overline{t} \le b.$$

Thus, for a given b, we can take

$$\overline{t} = \frac{b-1}{b^2}$$

by this argument and know that a solution $x(t)$ exists up to $t = \overline{t}$ and is bounded by b. The largest $\overline{t}$ we can get this way is obtained by choosing b to maximize $(b-1)/b^2$. To do this, put $\overline{t}(b) = (b-1)/b^2$, then

$$\frac{d\overline{t}(b)}{db} = \frac{1}{b^2} - \frac{2(b-1)}{b^3}$$

and setting $d\overline{t}(b)/db = 0$, we get the equation

$$\frac{1}{b^2} - \frac{2(b-1)}{b^3} = 0$$

which has the solution $b = 2$, giving us the information that a solution $x(t)$ bounded by $b = 2$ exists for all t up to $\overline{t} = (2-1)/2^2 = \frac{1}{4}$. The difference method can now be shown to converge for $T \le \frac{1}{4}$, using a few additional arguments.

At this point we could continue the solution beyond $t = \frac{1}{4}$, in this example, by constructing a new region ahead of the solution at $t = \frac{1}{4}$ proceeding as before. We could eventually reach any t for which the solution still exists; in this example that is any $t < 1$.

Even if one does not carry out a convergence analysis in a particular application of difference methods to the approximate solution of a differential equation problem, *at the very least* he should run the numerical computation with *more than one* discretization.

If the results for different step sizes (different numbers of grid points) are quite different as was the case in the example above with $T = 2$, then he should reject the numerical solutions as worthless. If there is good agreement, that is, at least, positive *evidence* for some kind of sensible result.

The *modified Euler method* is usually far more accurate than the Euler method for a given Δt and is again a step-by-step method; but there are two parts to the computation at each step:

1. Given an approximation $X(t_k)$ to the solution $x(t_k)$ at $t = t_k$, compute

$$X^{(1)}(t_{k+1}) = X(t_k) + \Delta t f[t_k, X(t_k)].$$

2. Compute, finally,

$$X(t_{k+1}) = X(t_k) + \Delta t \left[\frac{f[t_k, X(t_k)] + f[t_k, X^{(1)}(t_{k+1})]}{2} \right]$$

as an approximation to the solution $x(t_{k+1})$ at $t_{k+1} = t_k + \Delta t$.

In step (2) we are approximating the integral

$$\int_{t_k}^{t_{k+1}} f[t', x(t')] dt'$$

by the trapezoidal rule

$$(t_{k+1} - t_k) \frac{f[t_k, x(t_k)] + f[t_{k+1}, x(t_{k+1})]}{2};$$

except that we do not know $x(t_{k+1})$, so we approximate it using Euler's method in step (1).

EXERCISE

Repeat the previous exercise using the modified Euler method. That is, compute X_k, for $k = 10, 20, 40$, with $\Delta t = 0.5/k$ using

$$X_{j+1}^{(1)} = X_j + \Delta t X_j^2$$

$$X_{j+1} = X_j + \Delta t \frac{X_j^2 + X_{j+1}^{(1)2}}{2}$$

$(j = 0, 1, \ldots, k - 1)$ with $X_0 = 1$. Estimate $E_k = 2 - X_k$ as a function of Δt. Compare with Euler's method.

In both the Euler and modified Euler methods, the *step size* Δt can be varied from step-to-step if desired. Sometimes it is efficient to do so, using a larger step size when the solution is slowly varying than when it is rapidly varying.

Consider next the application of the modified Euler method to the differential equations of motion of a spaceship in orbit around the sun during the earth-Mars flight discussed in Appendix A.

We have

$$\ddot{x}_1 = \frac{-gx_1}{r^3}$$

$$\ddot{x}_2 = \frac{-gx_2}{r^3}$$

with $r = (x_1^2 + x_2^2)^{1/2}$. If we define auxiliary variables $x_3 = \dot{x}_1$ and $x_4 = \dot{x}_2$, then we can put the pair of *second order* (second derivatives) differential equations into the form of a system of four first-order differential equations, namely,

$$\dot{x}_1 = x_3$$
$$\dot{x}_2 = x_4$$
$$\dot{x}_3 = \ddot{x}_1 = \frac{-gx_1}{(x_1{}^2 + x_2{}^2)^{3/2}}$$
$$\dot{x}_4 = \ddot{x}_2 = \frac{-gx_2}{(x_1{}^2 + x_2{}^2)^{3/2}}.$$

These equations are in a form suitable for the application of the modified Euler method, with the vector function $f = (f_1, f_2, f_3, f_4)$ given by

$$f_1(t, x_1, x_2, x_3, x_4) = x_3$$
$$f_2(t, x_1, x_2, x_3, x_4) = x_4$$
$$f_3(t, x_1, x_2, x_3, x_4) = \frac{-gx_1}{(x_1{}^2 + x_2{}^2)^{3/2}}$$
$$f_4(t, x_1, x_2, x_3, x_4) = \frac{-gx_2}{(x_1{}^2 + x_2{}^2)^{3/2}}.$$

The first part of the computation at each step becomes, for this set of functions

(1)
$$X_1^{(1)}(t_{k+1}) = X_1(t_k) + \Delta t X_3(t_k)$$
$$X_2^{(1)}(t_{k+1}) = X_2(t_k) + \Delta t X_4(t_k)$$
$$X_3^{(1)}(t_{k+1}) = X_3(t_k) + \Delta t \left[\frac{-gX_1(t_k)}{[X_1{}^2(t_k) + X_2{}^2(t_k)]^{3/2}} \right]$$
$$X_4^{(1)}(t_{k+1}) = X_4(t_k) + \Delta t \left[\frac{-gX_2(t_k)}{[X_1{}^2(t_k) + X_2{}^2(t_k)]^{3/2}} \right]$$

and the second part of each step is

(2)
$$X_1(t_{k+1}) = X_1(t_k) + \frac{\Delta t}{2} [X_3(t_k) + X_3^{(1)}(t_{k+1})]$$

$$X_2(t_{k+1}) = X_2(t_k) + \frac{\Delta t}{2} [X_4(t_k) + X_4^{(1)}(t_{k+1})]$$

$$X_3(t_{k+1}) = X_3(t_k) + \frac{\Delta t}{2} \left[\frac{-gX_1(t_k)}{R_k{}^3} + \frac{-gX_1^{(1)}(t_{k+1})}{\bar{R}_{k+1}^3} \right]$$

$$X_4(t_{k+1}) = X_4(t_k) + \frac{\Delta t}{2} \left[\frac{-gX_2(t_k)}{R_k{}^3} + \frac{-gX_2^{(1)}(t_{k+1})}{\bar{R}_{k+1}^3} \right]$$

where

$$R_k = [X_1{}^2(t_k) + X_2{}^2(t_k)]^{1/2}$$

and

$$\bar{R}_{k+1} = \{[X_1^{(1)}(t_{k+1})]^2 + [X_2^{(1)}(t_{k+1})]^2\}^{1/2}.$$

Referring to the data, formulas, and descriptions of Sections 1.4, 5.4, and Appendix A, if we put

$$g = 4.15 \cdot 10^{17}$$
$$x_1(0) = 92.8 \cdot 10^6 \; (= \sqrt{E^2 - (5 \cdot 10^6)^2})$$
$$x_2(0) = 5 \cdot 10^6$$
$$x_3(0) = \dot{x}_1(0) = 0$$
$$x_4(0) = \dot{x}_2(0) = 73{,}430 \; (= V + V_0)$$

we obtain, in units of *miles* (for spatial coordinates x_1, x_2) and *hours* (for time), approximate initial values of position and velocity for a spaceship put into an orbit around the sun leaving the vicinity of the Earth so that the spaceship can reach the orbit of Mars (around the sun).

Using the modified Euler method, we can compute a step-by-step numerical solution to the differential equations of motion beginning with the given initial data.

We can make the numbers that will occur during the computation of more nearly uniform size for computing by a *change of scale*. If instead of miles, we use, say *10^8 miles* as the *unit of length* and if we use *10^3 hours* as the *unit of time*, then in these units, the initial data become

$$g = .415$$
$$x_1(0) = .928$$
$$x_2(0) = .05$$
$$x_3(0) = \dot{x}_1(0) = 0$$
$$x_4(0) = \dot{x}_2(0) = .7343.$$

A *rough* estimate of the time T (in units of 10^3 hours) required to reach the vicinity of Mars along the orbit chosen is $T = 6.0$.

EXERCISES

1. Using the units suggested, carry out (on a computing machine) the numerical solution just outlined. Plot some of the computed solution points $[X_1(t_k), X_2(t_k)]$ in order to sketch the orbit. Repeat the entire computation with $\Delta t = 0.1, 0.05, 0.025$. Compare results. Estimate the *time* when the computed orbit comes closest to the orbit of Mars and describe how to use that information to decide *when* to begin the flight in order to reach the vicinity of the planet Mars.

2. Using the transformation to polar coordinates given by

$$x_1 = r \cos \theta$$
$$x_2 = r \sin \theta,$$

derive the following equations of motion in polar coordinates:

$$\ddot{r} = \frac{-g}{r^2} + r\dot{\theta}^2$$

$$\ddot{\theta} = -\frac{2\dot{r}\dot{\theta}}{r}.$$

Write this system in an appropriate form for the application of the modified Euler method. Transform the initial data and repeat the numerical work of Exercise 1 but using polar coordinates. Compare with previous results. Discuss any advantages of using polar coordinates for this problem.

A beautiful treatment of the computational advantages (efficiency and accuracy) of transformations in problems in celestial mechanics can be found in *Linear and Regular Celestial Mechanics*, by E. L. Stiefel and G. Scheifele. New York: Springer-Verlag, 1971. See also Daniel and Moore (1970).

Suppose we seek a numerical solution $X(t)$ to an initial-valued problem for t in the range $t_0 \le t \le T$. Assume that a unique exact solution $x(t)$ exists for $t_0 \le t \le T$ for the particular $T > t_0$ of interest. If a step-by-step method is used with step size Δt such that $k\Delta t = T$, then we will reach $X(T)$ in k steps of size Δt. Denote this numerical solution at $t = T$ by $X(T,\Delta t)$. We define the *order* of the method as the largest integer p for which

$$\frac{X(T,\Delta t) - x(T)}{(\Delta t)^p}$$

is bounded as $\Delta t \to 0$ if infinite precision arithmetic is used. The definition of order assumes there is no round-off error.

Of course, the order of a method may depend on the particular differential equation to which it is applied. On the other hand, for a large *class* of equations it can be shown that Euler's method has order one and the modified Euler method is of order 2; see Henrici (1962). Thus, if $|X(T,\Delta t) - x(T)| < C(\Delta t)^p$, we have (at least) a pth-order method.

"High-order" methods (large p) are usually more efficient than low-order methods. In order to obtain a solution $X(T,\Delta t)$ of required accuracy we will have to compute k steps of size Δt, where Δt is sufficiently small. For a given method the amount of computation (number of arithmetic operations) will be proportional to the number of steps required. For a high-order method we can take a relatively (compared to a lower-order method) large Δt and so a small number of steps. On the other hand, the amount of computation *per step* will increase with increasing order of the method and so there will be a point of diminishing return. Theoretical and computational studies have shown that the optimal order to use to maximize efficiency is about the same as the number of decimal places of accuracy required; see Moore (1966 and 1968). For very low-accuracy requirements, low-order methods will often be most efficient. This will depend also on the range in t over which the solution is sought. Over very long ranges $[t_0, T]$ in the independent variable t, even low-accuracy requirements may be more difficult to meet and the optimal order will increase with T.

Many methods of high order have been derived (see Henrici, 1962; Babuska et al., 1966; and Daniel and Moore, 1970). There is extensive mathematical literature on various high-order methods in the journals, for example, *Numerische Mathematik*, *SIAM Journal of Numerical Analysis*, *Mathematics of Computation*, and so forth. For analytic differential equations, successive finite Taylor series expansions of arbitrarily high order may be used to give a method of any desired order.

There seems to be widespread misunderstanding concerning the efficiency of the Taylor series method. With the advent of computers with large storage capacity we can generate Taylor coefficients recursively in an efficient manner; see Section 6.2 and also Moore (1966) and Rall (1965, Vol. I, pp. 103–112, 185–203). Actually, in numerical solution of problems in celestial mechanics, Taylor series have been used in this way (with recursive evaluation of coefficients) for some time; see, for instance, Steffensen (1956).

To explain and illustrate the Taylor series method, consider first the equation

$$\frac{dx}{dt} = t^2 + x^2$$

(see P. Henrici, 1962, p. 66).

Using the notation of Section 6.2 we have

$$(x)_1 = t^2 + x^2 = (t)_0(t)_0 + (x)_0(x)_0$$

$$(x)_k = \frac{1}{k}\,[(x)_1]_{k-1} = \frac{1}{k}\sum_{j=0}^{k-1}\,[(t)_j(t)_{k-1-j} + (x)_j(x)_{k-1-j}]$$

$$k = 1, 2, \ldots .$$

Furthermore,

$$(t)_0 = t$$
$$(t)_1 = 1$$
$$(t)_k = 0 \quad \text{for } k > 1.$$

From this recursion formula for $(x)_k$ we can compute the Kth-order Taylor series expansion for the solution at $t + h$

$$x(t + h) = \sum_{k=0}^{K} (x)_k(t)h^k$$

from a given value of t and $x(t) = (x)_0$ and h and K. Beginning with a given initial value x_0 at $t = 0$, say, we can compute in this way successive approximate solution values at $t = t_i$, with $t_i = t_{i-1} + h_i$, $(i = 1, 2, \ldots)$, using the recursion formula to compute the coefficients $(x)_k(t_i)$ at $t = t_i$.

Let us examine more closely the special case that we choose the fourth order Taylor series method $(K = 4)$. The recursion formula for the Taylor coefficients of the solution of the differential equation

$$(x)_1 = \frac{dx}{dt} = t^2 + x^2$$

is

$$(x)_k = \frac{1}{k} \sum_{j=0}^{k-1} [(t)_j(t)_{k-1-j} + (x)_j(x)_{k-1-j}]$$

for $k = 2, 3, 4$, with $(t)_0 = t$, $(t)_1 = 1$, and $(t)_k = 0$ when $k = 2,3,4$.

We could also write out the formula explicitly for $k = 2, 3, 4$ as

$$(x)_2 = t + (x)_0(x)_1$$
$$(x)_3 = \tfrac{1}{3}[1 + 2(x)_0(x)_2 + (x)_1^2]$$
$$(x)_4 = \tfrac{1}{2}[(x)_0(x)_3 + (x)_1(x)_2].$$

In this form we can count the operations (multiplications and additions) required to compute one step of the fourth-order Taylor method for this example. We count 5 additions and 10 multiplications (storing $\tfrac{1}{3}$ and $\tfrac{1}{2}$) to compute $(x)_1$, $(x)_2$, $(x)_3$, and $(x)_4$ and 4 more multiplications and 4 additions to evaluate

$$x(t + h) = \sum_{k=0}^{4} (x)_k h^k$$
$$= h(h\{h[h(x)_4 + (x)_3] + (x)_2\} + (x)_1) + (x)_0.$$

Altogether, this makes 9 additions and 14 multiplications.

Let us compare this operation count with that of another, widely used, fourth-order method, called the *Runge-Kutta* method. For the differential equation

$$\frac{dx}{dt} = f(t,x)$$

The Runge-Kutta method approximates $x(t + h)$ in terms of $(x)_0 = x(t)$ by $x(t + h) = (x)_0 + \tfrac{1}{6}h(k_1 + 2(k_2 + k_3) + k_4)$ where

$$k_1 = f[t,(x)_0]$$
$$k_2 = f[t + \tfrac{1}{2}h,(x)_0 + \tfrac{1}{2}hk_1]$$
$$k_3 = f[t + \tfrac{1}{2}h,(x)_0 + \tfrac{1}{2}hk_2]$$
$$k_4 = f(t + h,(x)_0 + hk_3).$$

This method requires 4 evaluations of f plus 9 additions and 7 multiplications to evaluate $x(t + h)$ and the arguments of f which are needed. For the particular example

$$\frac{dx}{dt} = f(t,x) = t^2 + x^2$$

each evaluation of f requires 1 addition and 2 multiplications. Altogether, then the fourth-order Runge-Kutta method requires, for this example, 13 additions and 15 multiplications per step — just a little *more* than the fourth-order Taylor method (9 and 14).

EXERCISE

Derive a recursion formula for the Taylor coefficients of solutions of the differential equation

$$\frac{dx}{dt} = x^2.$$

Apply the eighth-order $(K = 8)$ Taylor method in a step-by-step fashion to obtain an approximate numerical value for $x(0.5)$ beginning with the initial value $x(0) = 1$. Do this for $h = 0.5$ in one step, for $h = 0.25$ in two steps, and for $h = 0.1$ in five steps. Explain how the *results* indicate that an eighth-order method is being used. (*Hint:* See Section 6.3.)

In a computational study (reported in Appendix C) the Newtonian equations of motion for the system of five gravitational bodies, sun, Mercury, Venus, earth, and Jupiter, were integrated numerically over a 100-year period of real time (for the century 1850–1950). This was carried out on the UW1108 computer in double-precision arithmetic (about 18 decimals) using the twelfth-order Taylor method. Step size was chosen to maintain the relative error per step* at about 10^{-12} for the position of the planet Mercury, which carried out about 415 orbits around the sun during the 100-year period. The ten known integrals to the N-body problem were printed out (along with the approximate solution) and these all remained constant to *at least* 8 decimals throughout. Total computation time was less than an hour on the UW1108. The program was written in FORTRAN V. A number of auxiliary computations were included for the determination of the position of successive perihelia of the planet Mercury for the purpose of reverifying a prediction of the general theory of relativity.

For the thirtieth-order system of differential equations involved in this calculation, the twelfth-order Taylor method required about five times as many arithmetic operations per step as the fourth-order Runge-Kutta

*As estimated by the last (12th) term in the finite Taylor series computed at each point.

method would have. On the other hand, a much smaller step size would have been required for comparable accuracy using the fourth-order Runge-Kutta method so that far more than five times as many steps would have been needed and the total computing time would have been many times greater.

An even more drastic comparison could be made with a second-order method such as the modified Euler method. Actual computation time was observed for a solution carried out part way using a second-order method. To complete the computation would have required several hundred hours on the UW1108.

Other high-order methods may be somewhat more efficient than the Taylor method on a given problem. The main point here is that low-order methods will seldom be most efficient.

For discussions of error analysis in numerical solutions of differential equations, see Babuska et al., 1966; Henrici, 1962; Moore, 1966; and Rall, 1965.

6.5 Two-Point, Boundary-Value Problems

It can be shown (Moore, 1960) that a two-point boundary value problem of the form

$$\frac{d^2y}{d\bar{t}^2} = g(\bar{t},y)$$

$$y(a) = y_1, \quad y(b) = y_2$$

has a (unique) solution in a region S containing the line segment connecting the points y_1 and y_2, provided that $g(\bar{t},y)$ and $(\partial g/\partial y)(\bar{t},y)$ are continuous for y in S and $\bar{t}$ in $[a,b]$ and provided that $b - a$ is sufficiently small. By an easy change of variables the problem can be put into the form

$$\frac{d^2x}{dt^2} = f(t,x)$$

$$x(0) = x(1) = 0.$$

Put

$$t_i = \frac{i}{N} \qquad (i = 0,1,2,\ldots,N)$$

and consider the following discretization:

$$x_{i+1} - 2x_i + x_{i-1} = \left(\frac{1}{N}\right)^2 f(t_i,x_i) \qquad (i = 1,2,\ldots,N-1)$$

where $x_0 = x_N = 0$. This is obtained by approximating the second derivative of x by the second difference quotient

$$\frac{d^2x}{dt^2} \approx \frac{x(t+h) - 2x(t) + x(t-h)}{h^2}$$

with $h = 1/N$. This discretization has the form

$$Mx = \left(\frac{1}{N}\right)^2 f$$

where

$$x = \begin{pmatrix} x_1 \\ x_2 \\ \cdot \\ \cdot \\ \cdot \\ x_{N-1} \end{pmatrix}, \quad f = \begin{pmatrix} f(t_1, x_1) \\ \cdot \\ \cdot \\ \cdot \\ f(t_{N-1}, x_{N-1}) \end{pmatrix}$$

and where the tridiagonal matrix M has the components

$$M_{ij} = \begin{cases} 0, & |j-i| > 1 \\ -2, & j = i \\ 1, & |j-i| = 1 \end{cases}$$

We could make an initial guess at x_i, say $x_i^{(0)} = 0$ ($i = 1, 2, \ldots, N = 1$), and then iteratively compute the vectors $x^{(p+1)}$, $p = 0, 1, 2, \ldots$ by solving at each iteration the linear algebraic system

$$Mx^{(p+1)} = \left(\frac{1}{N}\right)^2 f^{(p)}$$

where

$$f_i^{(p)} = f(t_i, x_i^{(p)}).$$

An alternative method is the following.

It is not hard to verify that the matrix $K = (1/N)M^{-1}$ has the components

$$K_{ij} = \begin{cases} \left(\dfrac{i}{N} - 1\right)\dfrac{j}{N}, & j \le i \\ \left(\dfrac{j}{N} - 1\right)\dfrac{i}{N}, & j > i \end{cases}$$

and therefore a mathematically equivalent system is

$$x = \left(\frac{1}{N}\right)^2 M^{-1} f$$

or

$$x = \frac{1}{N} Kf.$$

In component form, this becomes

$$x_i = \frac{1}{N} \sum_{j=1}^{N-1} K_{ij} f(t_j, x_j) \qquad (i = 1, 2, \ldots, N-1).$$

Here, again, we could make an initial guess at x_i, say $x_i^{(0)} = 0$ ($i = 1, 2, \ldots,$ $N-1$) and iteratively compute the vectors $x^{(p+1)}$, $p = 0, 1, 2, \ldots,$ by *evaluating*

$$x^{(p+1)} = \frac{1}{N} K f^{(p)}$$

where $f_i^{(p)} = f(t_i, x_i^{(p)})$ as before.

EXERCISE

1. Compare the computational effort required for one iteration in the two methods described, taking advantage of the tridiagonal form of the matrix M.

2. Carry out a computational study of both methods for the two-point, boundary-value problem (for the *nonlinear pendulum* equations, see Daniel and Moore, 1970).

$$\frac{d^2 y}{d\bar{t}^2} = -\sin y$$

$$y(0) = 0, \quad y(1) = 10.$$

The second method given can be viewed also as a discretization of the integral equation

$$x(t) = \int_0^1 K(t, t') f[t', x(t')] dt'$$

where

$$K(t, t') = \begin{cases} (t-1)t', & t' \le t \\ (t'-1)t, & t' > t \end{cases}$$

is called the Green's function for the differential operator d^2/dt^2 with the given boundary conditions $x(0) = x(1) = 0$.

For further study of numerical methods for two-point, boundary-value problems see Babuska et al., 1966; Collatz, 1960; Daniel and Moore, 1970.

There are special methods that may be used for *linear* boundary-value problems, for instance, using Fourier series or Fourier integrals. Consider the two-point, linear boundary-value problem

$$x'' - bx = g(t)$$

$$x(0) = 0, \quad x(T) = 0.$$

Let

$$\varphi_K = \sin \frac{\pi K t}{T};$$

then

$$\varphi_K(0) = \varphi_K(T) = 0$$

and

$$\varphi''_K = -\left(\frac{\pi K}{T}\right)^2 \varphi_K.$$

Define

$$(u,v) = \int_0^T u(t)v(t)dt;$$

then

$$(\varphi_K,\varphi_j) = \int_0^T \sin \frac{\pi K t}{T} \sin \frac{\pi j t}{T}\, dt = \begin{cases} 0, & K \neq j \\ \dfrac{T}{2}, & K = j \end{cases}$$

Put

$$g(t) = \sum_{K=1}^{\infty} c_K \varphi_K$$

$$x(t) = \sum_{K=1}^{\infty} a_K \varphi_K.$$

Now

$$\varphi''_K - b\varphi_K = -\left[b + \left(\frac{\pi K}{T}\right)^2\right]\varphi_K$$

so we have, formally, $x'' - bx = g(t)$, if

$$a_K = \frac{-c_K}{b + \left(\dfrac{\pi K}{T}\right)^2} \qquad K = 1, 2, \ldots$$

where

$$c_K = \frac{(g,\varphi_K)}{\|\varphi_K\|^2} = \frac{2}{T}\int_0^T g(t) \sin \frac{\pi K t}{T}\, dt.$$

In other words, the Fourier series

$$x(t) = \sum_{K=1}^{\infty} \left(\frac{-c_K}{b + \left(\dfrac{\pi K}{T}\right)^2}\right) \sin \frac{\pi K t}{T}$$

with

$$c_K = \frac{2}{T} \int_0^T g(t) \sin \frac{\pi K t}{T} \, dt$$

solves the boundary-value problem (if b is not equal to $-(\pi K/T)^2$ for any $K = 1, 2, \ldots$).

Now put $S_K = \pi K/T$, where $K = 1, 2, \ldots$. The points S_K are evenly spaced on $[0, \infty]$ with $\Delta S = S_{K+1} - S_K = \pi/T$. The solution just obtained can be written as

$$x(t) = \sum_{K=1}^{\infty} \frac{-c_K}{b + S_K{}^2} \sin S_K t$$

$$c_K = \frac{2}{T} \int_0^T f(t) \sin S_K t \, dt.$$

Now suppose that $c(s) \equiv \int_0^\infty g(t) \sin st \, dt$ exists for all s in $[0, \infty]$. Then

$$x(t) = \sum_{k=1}^{\infty} \left(\frac{-c(s_k)}{b + s_k{}^2} \sin s_k t \right) \frac{2}{T}$$

$$= -\frac{2}{\pi} \sum_{k=1}^{\infty} \left(\frac{c(s_k)}{b + s_k{}^2} \sin s_k t \right) \Delta s$$

and, as $T \to \infty$ and $\Delta s = (\pi/T) \to 0$, we obtain, *formally*, the Fourier integral

$$x(t) = -\frac{2}{\pi} \int_0^\infty \frac{c(s)}{b + s^2} \sin st \, ds$$

$$c(s) = \int_0^\infty g(t) \sin st \, dt$$

as the solution of the problem with one boundary point at infinity

$$x'' - bx = g(t)$$
$$x(0) = 0, \quad x(\infty) = 0;$$

that is, $x(T) \to 0$ as $T \to \infty$.

For example, suppose

$$g(t) = e^{-t};$$

we obtain, from the formal solution,

$$C(s) = \int_0^\infty e^{-t} \sin st \, dt = -e^{-t} \sin st \Big|_0^\infty + \int_0^\infty s e^{-t} \cos st \, dt$$

$$= s \left(-e^{-t} \cos st \Big|_0^\infty - \int_0^\infty s e^{-t} \sin st \, dt \right)$$

$$= s(1 - sC(s))$$

so

$$C(s) = \frac{s}{1 + s^2}$$

and

$$x(t) = -\frac{2}{\pi} \int_0^\infty \frac{s}{(1 + s^2)(b + s^2)} \sin st \, ds.$$

If $b = 1$, then

$$x(t) = -\frac{2}{\pi} \int_0^\infty \frac{s}{(1 + s^2)^2} \sin st \, ds$$

$$= -\frac{2}{\pi} \left\{ \left(\sin st \right) \left(\frac{-1}{2(1 + s^2)} \right) \Big|_0^\infty + \int_0^\infty \frac{t}{2(1 + s^2)} \cos st \, ds \right\}$$

$$= -\frac{t}{\pi} \int_0^\infty \frac{\cos st}{1 + s^2} \, ds$$

$$= -\frac{t}{\pi} \cdot \frac{\pi}{2} \cdot e^{-t}.$$

Thus $x(t) = -(t/2)e^{-t}$. We can verify that $x'' - x = e^{-t}$ and $x(0) = 0$
$x(\infty) = 0$.

6.6 Finite Difference Methods for Partial Differential Equations

Under certain conditions, electrostatic or hydrostatic potential functions can be assumed to satisfy *Laplace's equation* which in *two independent variables* is the *partial differential equation*

$$\frac{\partial^2 u(x,y)}{\partial x^2} + \frac{\partial^2 u(x,y)}{\partial y^2} = 0.$$

For instance, for an *irrotational* planar flow of an *ideal* fluid (assumed to be inviscid and incompressible), there exists a velocity potential function $u(x,y)$ satisfying Laplace's equation. The components of velocity of a point in the fluid are given by $v_x = \partial u/\partial x$ and $v_y = \partial u/\partial y$.

Laplace's equation can also be taken to represent a relation satisfied by the steady-state temperature distribution $u(x,y)$ over a planar region within boundaries along which a steady (time-independent) temperature profile is maintained.

Similarly, Laplace's equation may be assumed to be satisfied by the local, steady-state concentrate or *density* $u(x,y)$ at a point x, y of any substance (chemicals, pollutants, ions, and so forth), which is allowed to diffuse (or conduct) through some homogeneous medium over a planar region within boundaries along which some distribution of values of $u(x,y)$ is maintained.

For a three-dimensional version of Laplace's equation, we may use

$$\frac{\partial^2 u(x,y,z)}{\partial x^2} + \frac{\partial^2 u(x,y,z)}{\partial y^2} + \frac{\partial^2 u(x,y,z)}{\partial z^2} = 0.$$

We may assign *boundary conditions* on a rectangle, for instance, in the x, y plane by giving values of $u(x,y)$ for (x,y) on the boundary of the rectangle. Suppose we have the boundary data:

$$
\begin{aligned}
u(x,y) = \;& \varphi_1(y) \text{ for } x = 0; \quad 0 \le y < b \\
& \varphi_2(x) \text{ for } y = b; \quad 0 \le x < a \\
& \varphi_3(y) \text{ for } x = a; \quad 0 < y \le b \\
& \varphi_4(x) \text{ for } y = 0; \quad 0 < x \le a
\end{aligned}
$$

where φ_1, φ_2, φ_3, and φ_4 are given functions.

We can compute approximations to the solution values $u(x,y)$ at interior points of the rectangle by replacing the differential equation by an approximating difference equation, as we will now show. Figure 6.6 illustrates the *geometry* of the problem.

Subdividing $[0,a]$ into N parts and $[0,b]$ into M parts, we can put a *rectangular grid* (Figure 6.7) of points on the region shown by defining

$$x_i = \frac{ia}{N}, \qquad i = 0, 1, 2, \ldots, N$$

$$y_j = \frac{jb}{M}, \qquad j = 0, 1, 2, \ldots, M.$$

We can approximate partial derivatives of $u(x,y)$ at a point (x_i,y_j) in the grid by difference quotients using values of u at *neighboring* points.

Using the notation

$$u_{i,j} = u(x_i,y_j), \quad \left(\frac{\partial u}{\partial x}\right)_{i,j} = \frac{\partial u(x_i,y_j)}{\partial x}, \cdots$$

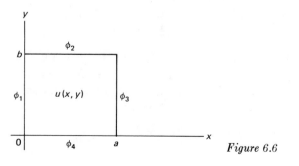

Figure 6.6

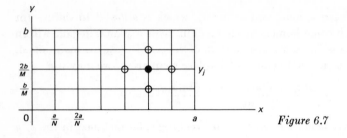

Figure 6.7

we can approximate partial derivatives at (x_i, y_j) as follows:

$$\left(\frac{\partial u}{\partial x}\right)_{i,j} \approx \frac{u_{i+1,j} - u_{i,j}}{\Delta x}, \quad \text{where } \Delta x = \frac{a}{N}$$

$$\left(\frac{\partial u}{\partial y}\right)_{i,j} \approx \frac{u_{i,j+1} - u_{i,j}}{\Delta y}, \quad \text{where } \Delta y = \frac{b}{M}$$

$$\left(\frac{\partial^2 u}{\partial x^2}\right)_{i,j} \approx \frac{\left(\frac{\partial u}{\partial x}\right)_{i,j} - \left(\frac{\partial u}{\partial x}\right)_{i-1,j}}{\Delta x} \approx \frac{u_{i+1,j} - 2u_{i,j} + u_{i-1,j}}{(\Delta x)^2}$$

$$\left(\frac{\partial^2 u}{\partial y^2}\right)_{i,j} \approx \frac{u_{i,j+1} - 2u_{i,j} + u_{i,j-1}}{(\Delta y)^2}.$$

With these difference quotients we can approximate Laplace's equation by the difference equation

$$\frac{u_{i+1,j} - 2u_{i,j} + u_{i-1,j}}{(\Delta x)^2} + \frac{u_{i,j+1} - 2u_{i,j} + u_{i,j-1}}{(\Delta y)^2} = 0.$$

We can use this difference equation to compute an approximate solution to a given boundary-value problem of the form described. To do this, we want to require that the difference equation be satisfied by the computed values of the solution at all interior points of the rectangle.

We can give the interior grid points a *linear ordering*, for instance, as follows:

$$\text{denote the point } (x_i, y_j) \text{ by}$$
$$P_{i+(j-1)(N-1)} = (x_i, y_j) \qquad (i = 1, 2, \ldots, N-1),$$
$$(j = 1, 2, \ldots, M-1);$$

denote by U_k the sought-after numerical solution at P_k [$k = 1, 2, \ldots, (M-1)(N-1)$]. For a given k, we have $i \equiv k[\text{mod } (N-1)]$ and $j = (k-i)/(N-1) + 1$. Using this notation, the difference equation at interior points P_k not bordering the boundary becomes

$$\frac{U_{k+1} - 2U_k + U_{k-1}}{(\Delta x)^2} + \frac{U_{k+N-1} - 2U_k + U_{k-N+1}}{\Delta y^2} = 0.$$

Let

$$c_1 = -\frac{1}{2}\left[1 + \frac{(\Delta y)^2}{(\Delta x)^2}\right]^{-1}$$

and

$$c_2 = -\frac{1}{2}\left[1 + \frac{(\Delta x)^2}{(\Delta y)^2}\right]^{-1}$$

then the difference equation at points P_k not bordering the boundary can be written

$$c_1 U_{k-N+1} + c_2 U_{k-1} + U_k + c_2 U_{k+1} + c_1 U_{k+N-1} = 0.$$

If $\Delta x = a/N$ and $\Delta y = b/M$ are about the same size, then c_1 and c_2 are near $-\frac{1}{4}$.

For points that border the boundary some of these terms will be given as boundary conditions. Such terms can be put on the right-hand side of the difference equations for those points. Putting these equations together we have a system of linear algebraic equations to solve. Using matrix and column vector notation, we seek a column vector U (with components $U_1, U_2, \ldots, U_{(M-1)(N-1)}$) which satisfies

$$CU = \Phi$$

where C is the matrix whose elements are all zero except for five or fewer entries per row. The diagonal elements $C_{k,k}$ are all 1.

We can put

$$C = I + M,$$

then we can write the system as

$$U = \Phi - MU$$

where the diagonal elements of M are all zero.

If we choose an initial approximation $U^{(0)}$, we can use the *Gauss-Seidel method* in which we replace each component of $U^{(p)}$ by its *new* value as we run through the components of $U^{(p+1)}$. Thus, we compute

$$U_k^{(p+1)} = \Phi_k - \sum_{j=1}^{k-1} M_{kj} U_j^{(p+1)} - \sum_{j=k+1}^{(M-1)(N-1)} M_{kj} U_j^{(p)}$$

for $k = 1, 2, \ldots, (M-1)(N-1)$, and we can continue this for $p = 0, 1, 2, \ldots$ until some stopping criterion is met, for instance, until $p = P$, where P is chosen in advance. For a thorough treatment of finite difference methods for partial differential equations, see, for example, Forsythe and Wasow (1960).

We may try to solve Laplace's equation on a three-dimensional rectangular region; if we discretize with only 10 points on an edge we will already have a linear system of order 1000 to solve. *Large* linear systems thus arise easily and naturally in finite difference methods applied to partial differential equations (see also Young, 1971; Varga, 1962).

Asymptotic behavior of solutions

7.1 Introduction

Often, a time-dependent process, once set into motion, will settle down to a simple *steady-state* behavior after a short-lived ("transient") behavior of a more complicated nature. In this chapter we consider three examples of this kind of "asymptotic" behavior. For each example, we discuss methods for deriving the nature of the steady-state solution and methods for estimating the *time* required to essentially reach the steady-state solution.

7.2 Terminal Velocity of Fall in a Resisting Medium

When an object is subjected to a constant accelerating force in a resisting medium, the friction or *drag* of the medium (air or water, say) will cause the velocity of the object to tend toward a certain limiting or *terminal* velocity. For instance, consider an object falling in the earth's atmosphere. The downward acceleration of gravity acting on the falling object is inversely proportional in magnitude to the square of the distance from the object to the center of the earth. This is a continuous function of the distance and so over a short range of distances it may be approximated by a constant. Over a four-mile fall to the surface of the earth, the acceleration of gravity acting upon a falling object will vary by about 0.2 percent. To a first approximation we may take it to be constant, say $-g$. At subsonic speeds and at low altitudes (under a few miles) it can be determined experimentally that the resisting force of drag provided upon a falling object by the earth's atmosphere is approximately proportional to the square of the velocity of the falling object and is also dependent on the area of the object (larger for instance, with the parachute *open*). Thus a reasonable model for a *law of motion* for an object falling through the atmosphere under the conditions stated (subsonic velocities and low altitudes) is given by the differential equation

$$\frac{dv}{dt} = -g + cv^2$$

where v is the velocity of the falling object measured as *positive* going away from the earth. For a *falling* object, v will be *negative*. The constant c depends on the particular object and can be found experimentally.

EXERCISE

Describe some experimental means for determining a value for c for a given object.

Consider the differential equation just written. For a positive constant c, we can see directly from the equation that when v^2 is less than g/c, then the fall will be getting faster (v negative and dv/dt negative). When v^2 is greater than g/c, then the fall will be slowing (v negative and dv/dt positive). When v^2 is equal to g/c, then dv/dt is zero and the object continues to fall at the constant negative velocity: $-\sqrt{(g/c)}$. According to the differential equation, the velocity v *tends* toward and then stays at the steady-state constant velocity: $-\sqrt{(g/c)}$.

The *time* required for v to get within, say, 99 percent of its steady-state value beginning with $v = v_0$ at time zero, can be estimated roughly using Euler's method.

We put

$$v(t + \Delta t) \approx v(t) + \Delta t \frac{dv(t)}{dt}.$$

For $v(0) = v_0$ and $v(\Delta t) = .99(-\sqrt{g/c})$, we have approximately

$$.99(-\sqrt{g/c}) = v_0 + \Delta t(-g + cv_0^2)$$

or

$$\Delta t = \frac{.99(-\sqrt{g/c}) - v_0}{-g + cv_0^2}.$$

EXERCISE

Describe a procedure for computing a more accurate estimate of the time
required for $v(t)$ to reach within .99 of its steady-state value.

7.3   Steady-State Response of an Antenna Circuit

The antenna circuit of a radio receiver contains an inductor and a variable
capacitor in series. An AM radio transmission with a carrier frequency f will
induce a steady-state alternating current with frequency f within the antenna circuit after a *transient* current dies out exponentially. The amplitude
of the induced alternating current will depend on the setting of the variable
capacitor. And when the setting makes the circuit *tuned* to the frequency f,
then the amplitude will be maximum. In this setting, the circuit is said to be
in *resonance* for the frequency f. A simple diagram will illustrate the relevant
features of the type of circuit under discussion; see Figure 7.1.

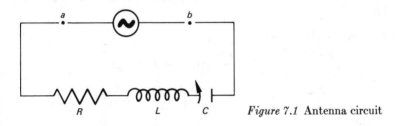

Figure 7.1 Antenna circuit

We suppose that the transmitted radio signal induces a sinusoidal alternating potential difference across the terminals a, b. Then the current $I(t)$
in the circuit as a function of time t satisfies the differential equation

$$L \frac{d^2I(t)}{dt^2} + R \frac{dI(t)}{dt} + \frac{1}{C} I(t) = 2\pi f V_m \cos 2\pi ft$$

where L, R, and C are, respectively, the inductance, resistance, and capacitance of the circuit and V_m is the amplitude of the induced potential difference across the terminals. It can be verified, by differentiating and substituting in the differential equation, that

$$I(t) = K \sin (2\pi ft - \phi)$$

satisfies the differential equation provided K and ϕ are chosen so that

$$K = \frac{V_m}{\sqrt{R^2 + \left(2\pi fL - \frac{1}{2\pi fC}\right)^2}}$$

and

$$\tan \phi = \frac{2\pi fL - \dfrac{1}{2\pi fC}}{R}.$$

This *is* the *steady-state* solution of the differential equation as will be shown presently. We can observe that, for a given resistance R and inductance L, the amplitude K will be *maximum* when the capacitance C is set to the value $C = 1/(2\pi f)^2 L$. For this reason, the tuned frequency

$$f = \frac{1}{2\pi \sqrt{LC}}$$

is called the *natural-resonant frequency* of the circuit. For a particular setting of C, the circuit will be in *resonance* for the frequency $1/2\pi \sqrt{LC}$; that is, the *maximum amplitude of response* will be for that frequency. If, at the *onset* of a transmitted signal, say at time $t = 0$, the current in the circuit is $I(0)$ and the *instantaneous rate of change* of current is $(dI/dt)(0)$, then the exact solution $I(t)$ to the differential equation for those initial conditions can be expressed as

$$I(t) = K \sin (2\pi ft - \phi)$$
$$+ \left(\frac{\alpha_1 e^{-(\alpha_2 t/L)} - \alpha_2 e^{-(\alpha_1 t/L)}}{\alpha_1 - \alpha_2} \right)(I(0) + K \sin \phi)$$
$$+ \frac{L}{\alpha_1 - \alpha_2} \left(e^{-(\alpha_2 t/L)} - e^{-(\alpha_1 t/L)} \right)\left(\frac{dI(0)}{dt} - K2\pi f \cos \phi \right)$$

with

$$K = \frac{Vm}{\sqrt{R^2 + \left(2\pi fL - \dfrac{1}{2\pi fC} \right)^2}}$$

where

$$\phi = \tan^{-1} \left(\frac{2\pi fL - \dfrac{1}{2\pi fC}}{R} \right)$$

and

$$\alpha_1 = \frac{R}{2} + \frac{1}{2}\sqrt{R^2 - \frac{4L}{C}}$$

and

$$\alpha_2 = \frac{R}{2} - \frac{1}{2}\sqrt{R^2 - \frac{4L}{C}}.$$

The dependence of the current $I(t)$ upon the particular initial values $I(0)$ and $dI(0)/dt$ is *transient* and will become negligible as soon as t is large enough so that

$$e^{-(\alpha_1 t/L)} \quad \text{and} \quad e^{-(\alpha_2 t/L)}$$

are small. It remains to *derive* the expression just given for $I(t)$. The technique that will be used is widely applicable to many other equations of the same type and so a *general* discussion will be given first and then the equation under discussion will be solved by *specializing* the general technique.

To put the equation into a more general setting, call $I(t) = x_1(t)$ and $dI(t)/dt = x_2(t)$, then we can write the differential equation for $I(t)$ as

$$L \frac{dx_2(t)}{dt} + Rx_2(t) + \frac{1}{C} x_1(t) = 2\pi f V_m \cos 2\pi f t.$$

We also have

$$\frac{dx_1(t)}{dt} = x_2(t).$$

Thus we can write the equation in the form of a *vector differential equation*

$$\frac{dx(t)}{dt} = Ax(t) + F(t)$$

where

$$x(t) = \begin{pmatrix} x_1(t) \\ x_2(t) \end{pmatrix}$$

and A is the *matrix of constant coefficients*

$$A = \begin{pmatrix} 0 & 1 \\ -\dfrac{1}{LC} & -\dfrac{R}{L} \end{pmatrix}$$

and $F(t)$ is the vector function

$$F(t) = \begin{pmatrix} 0 \\ \dfrac{2\pi f V_m \cos 2\pi f t}{L} \end{pmatrix}.$$

We can carry out much of the discussion of the vector differential equation

$$\frac{dx(t)}{dt} = Ax(t) + F(t)$$

and its solutions in terms of properties of the constant matrix A. We will do this now for an arbitrary $n \times n$ matrix A and then apply the results to the special case at hand. We are interested in the nature of the solutions to the equation just written when A is an $n \times n$ matrix of real numbers and $F(t)$ is an n-dimensional vector

$$F(t) = \begin{pmatrix} F_1(t) \\ \cdot \\ \cdot \\ \cdot \\ F_n(t) \end{pmatrix}$$

whose components $F_1, \ldots, F_n$ are given piecewise continuous functions defined for all t.

We can define the *exponential* of any matrix A as the *matrix* (where I is the identity matrix),

$$e^A = I + A + \frac{A^2}{2!} + \frac{A^3}{3!} + \cdots.$$

Now At is a matrix for each real number t and

$$e^{At} = I + At + \frac{A^2 t^2}{2!} + \frac{A^3 t^3}{3!} + \cdots;$$

furthermore

$$\frac{d}{dt} e^{At} = A + A^2 t + \frac{A^3 t^2}{2!} + \cdots$$

$$= A \left(I + At + \frac{A^2 t^2}{2!} + \cdots \right)$$

$$= A e^{At}.$$

If we put

$$x(t) = e^{At} u(t),$$

then

$$\frac{dx(t)}{dt} = e^{At} \frac{du(t)}{dt} + A e^{At} u(t)$$

$$= e^{At} \frac{du(t)}{dt} + A x(t).$$

If we require that $u(t)$ satisfy

$$e^{At} \frac{du(t)}{dt} = F(t),$$

then we will have $x(t) = e^{At} u(t)$ as a solution to

$$\frac{dx(t)}{dt} = F(t) + A x(t).$$

We can write an expression for $u(t)$ as

$$u(t) = u(0) + \int_0^t e^{-At'} F(t') dt'.$$

We can verify, by differentiation, that

$$\frac{du(t)}{dt} = e^{-At} F(t)$$

and so

$$e^{At} \frac{du(t)}{dt} = e^{At} e^{-At} F(t)$$
$$= I F(t)$$
$$= F(t)$$

EXERCISE

Show that $e^{At} \cdot e^{-At} = I$.

Now when $t = 0$, $x(0) = e^{A0} u(0)$; but $e^{A0} = e^0 = I$, so $x(0) = Iu(0) = u(0)$. Putting all this together, we obtain the expression

$$x(t) = e^{At} \left[x(0) + \int_0^t e^{-At'} F(t') dt' \right]$$

as the vector-function solution to the vector-differential equation

$$\frac{dx(t)}{dt} = Ax(t) + F(t)$$

with the vector of initial values $x(t) = x(0)$ at $t = 0$.

Under certain conditions on the matrix A (namely, the real parts of the *proper values* or *eigenvalues* of A are all negative), the term $e^{At} x(0)$ will have components that all *decay* exponentially and so represent a *transient* contribution to the solution. In this case, the term

$$e^{At} \int_0^t e^{-At'} F(t') dt'$$

then represents the steady-state solution. In particular, it is independent of $x(0)$, the vector of initial values.

In order to analyze the effect of multiplying the initial vector $x(0)$ by the matrix e^{At} and thereby deduce the behavior of the term $e^{At} x(0)$ in the solution $x(t)$ to the vector-differential equation $dx(t)/dt = Ax(t) + F(t)$, we need a more *transparent* representation of e^{At} than that given by the infinite series

$$e^{At} = I + At + \frac{A^2 t^2}{2!} + \frac{A^3 t^3}{3!} + \cdots.$$

Actually, the *powers* of a matrix A are not independent indefinitely. An $n \times n$ matrix A represents a linear transformation of E^n and for any vector x in E^n the $n + 1$ vectors Ix, Ax, A^2x, ..., A^nx are *linearly dependent;* that is, there are real numbers c_0, c_1, ..., c_n, not all zero, such that $c_0x + c_1Ax + c_2A^2x + \cdots + c_nA^nx = 0$. This is so, because *any* $n + 1$ vectors in E^n are linearly dependent.

EXERCISE

Why?

It turns out that *for any real $n \times n$ matrix A there are real constants c_0, c_1, ..., c_n, not all zero, such that*

$$(c_0I + c_1A + c_2A^2 + \cdots + c_nA^n)x = 0$$

for every x in E^n (Cayley-Hamilton theorem). This means that every power of A can be expressed as a linear combination of I, A, ..., A^n, because the only matrix B that makes $Bx = 0$ for every x in E^n is the *zero matrix* with $B_{ij} = 0$ for $i, j = 1, 2, ..., n$. In fact,

$$c_0I + c_1A + \cdots + c_nA^n = 0$$

means that

$$A^n = -\frac{1}{c_n}(c_0I + c_1A + c_2A^2 + \cdots + c_{n-1}A^{n-1});$$

or

$$A^m = -\frac{1}{c_m}(c_0I + c_1A + \cdots + c_{m-1}A^{m-1})$$

if

$$c_n = c_{n-1} = \cdots = c_{m+1} = 0.$$

If $m \leq n$ and

$$A^m = -\frac{1}{c_m}(c_0I + c_1A + \cdots + c_{m-1}A^{m-1}),$$

then

$$A^{m+1} = -\frac{1}{c_m}(c_0A + c_1A^2 + \cdots + c_{m-1}A^m)$$

$$= -\frac{1}{c_m}\left\{c_0A + c_1A^2 + \cdots + c_{m-2}A^{m-1}\right.$$

$$\left. + c_{m-1}\left[-\frac{1}{c_m}(c_0I + c_1A + \cdots + c_{m-1}A^{m-1})\right]\right\}.$$

We can express A^{m+k} as a linear combination of $I, A, \ldots, A^{m-1}$ by making repeated use of the reduction of A^m to a sum of lower powers of A.

If

$$(c_0 I + c_1 A + \cdots + c_n A^n)x = 0$$

for every x in E^n, then

$$p(A) = \sum_{k=0}^{n} c_k A^k \qquad \text{(with } A^0 = I)$$

is called the *characteristic polynomial* of A.

Actually, more often the characteristic polynomial of a matrix A is defined as

$$p(\lambda) = \sum_{k=0}^{n} c_k \lambda^k$$

with the same coefficients $c_0, c_1, \ldots, c_n$ as before, but in terms of a *complex variable* λ.

In any case, the polynomial $p(\lambda)$ has n complex zeros (counting multiplicities) and can be written, in *factored form* as $p(\lambda) = c_n(\lambda - \lambda_1)(\lambda - \lambda_2) \cdots (\lambda - \lambda_n)$ where $\lambda_1, \lambda_2, \ldots, \lambda_n$ are the (not necessarily distinct) zeros of $p(\lambda)$. These zeros of $p(\lambda)$ are called the characteristic values (or *proper values* or *eigenvalues*) of the matrix A.

The corresponding *factored form* of $p(A)$ would be

$$p(A) = c_n(A - \lambda_1 I)(A - \lambda_2 I) \cdots (A - \lambda_n I).$$

We know that $p(A)x = 0$ for every x in E^n and if *one* of the factors of $p(A)$ say $A - \lambda_k I$ *annihilates* a nonzero vector x, that is if $(A - \lambda_k I)x = 0$, then we say that x is a *characteristic vector* (or *proper vector* or *eigenvector*) of A. We then have

$$Ax = \lambda_k x \quad \text{with } x \neq 0$$

and

$$A^2 x = A(Ax) = A(\lambda_k x) = \lambda_k(Ax) = \lambda_k^2 x,$$

and

$$A^m x = \lambda_k^m x \qquad (m = 1,2,3,\ldots).$$

There are several ways of finding the coefficients of the characteristic polynomial $p(A)$. One approach is to seek the linear factors directly. Consider the condition that $A - \lambda I$ be a linear factor of $p(A)$. This will be the case if there is a nonzero vector x such that $(A - \lambda I)x = 0$. This, in turn, will be the case, if $A - \lambda I$ is a *singular* matrix. Thus given A, we can seek λ

such that $A - \lambda I$ is singular. This can be done in various ways. Often, the *determinant* of $A - \lambda I$ is considered. If the determinant of $A - \lambda I$ is zero, then $A - \lambda I$ is singular. The "determinant" of an $n \times n$ matrix A is the *volume* of the set

$$\left\{ Ax \mid 0 \leq x_i \leq 1, \, i = 1, 2, \ldots, n, \right\}$$

(which is the image, under the linear transformation represented by A, of the unit n-cube). Alternatively, we can consider, directly, the system of equations

$$(A_{11} - \lambda)x_1 + A_{12}x_2 + \cdots + A_{1n}x_n = 0$$
$$A_{21}x_1 + (A_{22} - \lambda)x_2 + \cdots + A_{2n}x_n = 0$$
$$\vdots$$
$$A_{n1}x_1 + A_{n2}x_2 + \cdots + (A_{nn} - \lambda)x_n = 0$$
$$x_1^2 + \quad x_2^2 + \cdots + x_n^2 - 1 = 0.$$

This is a system of $n + 1$ nonlinear equations in the variables $x_1, x_2, \ldots, x_n$, and λ. The last equation expresses the requirement that x be a nonzero vector. If x is any nonzero vector that satisfies $(A - \lambda I)x = 0$, then any scalar multiple of x, say ax, also satisfies $(A - \lambda I)(ax) = 0$. Therefore we may impose the condition that x be a nonzero vector by requiring that $\|x\| = 1$, which is what the last equation in the system does. In other words, if x is a solution of $(A - \lambda I)x = 0$, then so is $x/\|x\|$. We could try Newton's method, for instance, to solve the system for $x_1, x_2, \ldots, x_n$, and λ.

The subject of numerical determination of characteristic values (or *eigenvalues*) of matrices is thoroughly treated in the important work of Wilkinson, 1965. The discussion of eigenvalue problems given here is intended only to illustrate an important and common source of such problems in scientific computing and something of the mathematical nature of such problems.

Rather than go straight for the linear factors (for the characteristic values and characteristic vectors), we could go after the coefficients of the characteristic polynomial directly. From there we can perhaps find the characteristic values $\lambda_1, \lambda_2, \ldots, \lambda_n$ and can then find corresponding characteristic vectors by solving linear algebraic systems (see, however, Wilkinson, 1965).

It is important to note that the characteristic values of a *real* matrix A may be *complex numbers*, and that the *characteristic vectors* may have *complex components* as well.

Suppose $\lambda_1, \lambda_2, \ldots, \lambda_n$ are the characteristic values of A and suppose that $z_1, z_2, \ldots, z_n$ are characteristic vectors of A with

$$Az_k = \lambda_k z_k \quad \text{and} \quad z_k \neq 0.$$

Let Z be the $n \times n$ matrix whose columns are the column vectors $z_1, z_2, \ldots, z_n$. Then

$$AZ = A \begin{pmatrix} \cdot & \cdot & & \cdot \\ \cdot & \cdot & & \cdot \\ \cdot & \cdot & & \cdot \\ z_1 & z_2 & \cdots & z_n \\ \cdot & \cdot & & \cdot \\ \cdot & \cdot & & \cdot \\ \cdot & \cdot & & \cdot \end{pmatrix}$$

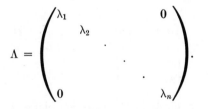

$$= \begin{pmatrix} \cdot & \cdot & & \cdot \\ \cdot & \cdot & & \cdot \\ \cdot & \cdot & & \cdot \\ z_1 & z_2 & \cdots & z_n \\ \cdot & \cdot & & \cdot \\ \cdot & \cdot & & \cdot \\ \cdot & \cdot & & \cdot \end{pmatrix} \begin{pmatrix} \lambda_1 & & & 0 \\ & \lambda_2 & & \\ & & \cdot & \\ & & & \cdot \\ 0 & & & \lambda_n \end{pmatrix} = Z\Lambda$$

where

$$\Lambda = \begin{pmatrix} \lambda_1 & & & 0 \\ & \lambda_2 & & \\ & & \cdot & \\ & & & \cdot \\ 0 & & & \lambda_n \end{pmatrix}.$$

Suppose Z is nonsingular, that is, $z_1, z_2, \ldots, z_n$ are linearly independent and thus form a *basis* for E^n (this will be the case, for instance, if $\lambda_1, \lambda_2, \ldots, \lambda_n$ are all different); then $A = Z\Lambda Z^{-1}$ and

$$A^2 = (Z\Lambda Z^{-1})(Z\Lambda Z^{-1})$$
$$= (Z\Lambda(Z^{-1}Z)\Lambda Z^{-1})$$
$$= Z\Lambda^2 Z^{-1}$$

and

$$A^m = Z\Lambda^m Z^{-1}, \quad \text{for } m = 1, 2, \ldots.$$

Now

$$\Lambda^m = \begin{pmatrix} \lambda_1{}^m & & & 0 \\ & \lambda_2{}^m & & \\ & & \cdot & \\ & & & \cdot \\ 0 & & & \lambda_n{}^m \end{pmatrix}$$

and so, for $A = Z\Lambda Z^{-1}$,

$$e^{At} = I + At + \frac{A^2 t^2}{2!} + \cdots$$

$$= ZIZ^{-1} + Z\Lambda t Z^{-1} + Z\frac{\Lambda^2 t^2}{2!} Z^{-1} + \cdots$$

$$= Ze^{\Lambda t} Z^{-1}$$

$$= Z\begin{pmatrix} e^{\lambda_1 t} & & & & 0 \\ & e^{\lambda_2 t} & & & \\ & & \cdot & & \\ & & & \cdot & \\ 0 & & & & e^{\lambda_n t} \end{pmatrix} Z^{-1}.$$

In this representation of e^{At} we can *see* that e^{At} is going to tend to the zero matrix, if all the quantities $e^{\lambda_1 t}, e^{\lambda_2 t}, \ldots, e^{\lambda_n t}$ tend toward zero.

If $\lambda_k = u + iv$ (with $i^2 = -1$), then

$$e^{\lambda_k t} = e^{ut + ivt} = e^{ut} \cdot e^{ivt}$$

with u, v, and t real numbers. Now

$$e^{ivt} = 1 + ivt + \frac{(ivt)^2}{2!} + \frac{(ivt)^3}{3!} + \cdots = 1 - \frac{(vt)^2}{2!} + \frac{(vt)^4}{4!} + \cdots$$

$$+ i\left(vt - \frac{(vt)^3}{3!} + \cdots\right) = (\cos vt + i \sin vt)$$

is a point on the unit circle in the complex plane. The term e^{ut} will tend toward zero if u is negative.

We can apply these notions to the study of the asymptotic behavior of the solution of the antenna-circuit equation. The matrix A, in that example, was

$$A = \begin{pmatrix} 0 & 1 \\ -\dfrac{1}{LC} & -\dfrac{R}{L} \end{pmatrix}.$$

Now

$$A^2 = \begin{pmatrix} -\dfrac{1}{LC} & -\dfrac{R}{L} \\ \dfrac{R}{L^2 C} & \left(-\dfrac{1}{LC} + \dfrac{R^2}{L^2}\right) \end{pmatrix},$$

so there will be constants c_0, c_1, and c_2 *not all zero* such that

$$(c_0 I + c_1 A + c_2 A^2)x = 0$$

for every vector

$$x = \begin{pmatrix} x_1 \\ x_2 \end{pmatrix};$$

that is, such that $c_0 I + c_1 A + c_2 A^2$ is the *zero matrix*, the 2×2 matrix

$$\begin{pmatrix} 0 & 0 \\ 0 & 0 \end{pmatrix}.$$

We can find c_0, c_1, c_2 directly, as follows. It is easy to determine that c_2 cannot be zero for the particular A we have since $c_0 I + c_1 A = 0$ would imply

$$c_0 \begin{pmatrix} 1 & 0 \\ 0 & 1 \end{pmatrix} + c_1 \begin{pmatrix} 0 & 1 \\ -\dfrac{1}{LC} & -\dfrac{R}{L} \end{pmatrix} = \begin{pmatrix} 0 & 0 \\ 0 & 0 \end{pmatrix}$$

and this would force $c_0 = c_1 = 0$. But we seek c_0, c_1, c_2 *not all zero*. We may as well put $c_2 = 1$, in this case. We then have the following equation for c_0 and c_1:

$$c_0 \begin{pmatrix} 1 & 0 \\ 0 & 1 \end{pmatrix} + c_1 \begin{pmatrix} 0 & 1 \\ -\dfrac{1}{LC} & -\dfrac{R}{L} \end{pmatrix} + \begin{pmatrix} -\dfrac{1}{LC} & -\dfrac{R}{L} \\ \dfrac{R}{L^2 C} & \left(-\dfrac{R}{LC} + \dfrac{R^2}{L^2} \right) \end{pmatrix} = \begin{pmatrix} 0 & 0 \\ 0 & 0 \end{pmatrix}.$$

Matching each of the four coefficients in the matrices, we must have c_0 and c_1 such that

$$c_0 \cdot 1 + c_1 \cdot 0 - \dfrac{1}{LC} = 0$$

$$c_0 \cdot 0 + c_1 \cdot 1 - \dfrac{R}{L} = 0$$

$$c_0 \cdot 0 + c_1 \cdot \left(-\dfrac{1}{LC} \right) + \dfrac{R}{L^2 C} = 0$$

$$c_0 \cdot 1 + c_1 \cdot \left(-\dfrac{R}{L} \right) + \left(-\dfrac{1}{LC} + \dfrac{R^2}{L^2} \right) = 0.$$

The first two equations are enough, and we find that

$$c_0 = \dfrac{1}{LC}$$

$$c_1 = \dfrac{R}{L}.$$

The third and fourth equations are then automatically also satisfied. Thus the characteristic polynomial for

$$A = \begin{pmatrix} 0 & 1 \\ -\dfrac{1}{LC} & -\dfrac{R}{L} \end{pmatrix}$$

is

$$p(A) = \frac{1}{LC} I + \frac{R}{L} A + A^2,$$

or, if we let λ be a complex variable, then

$$p(\lambda) = \frac{1}{LC} + \frac{R}{L} \lambda + \lambda^2.$$

The zeros of $p(\lambda)$ are

$$\lambda_1 = -\frac{R}{2L} + \frac{1}{2} \sqrt{\frac{R^2}{L^2} - \frac{4}{LC}}$$

$$\lambda_2 = -\frac{R}{2L} - \frac{1}{2} \sqrt{\frac{R^2}{L^2} - \frac{4}{LC}}.$$

EXERCISES

1. Find a characteristic vector z_1 "belonging to" λ_1 and a characteristic vector z_2 belonging to λ_2, that is z_1, z_2 such that

$$A z_1 = \lambda_1 z_1, \qquad z_1 \neq \begin{pmatrix} 0 \\ 0 \end{pmatrix}$$

$$A z_2 = \lambda_2 z_2, \qquad z_2 \neq \begin{pmatrix} 0 \\ 0 \end{pmatrix}.$$

2. Find the characteristic values of A if

$$A = \begin{pmatrix} 13 & -3 & 5 \\ 0 & 4 & 0 \\ -15 & 9 & -7 \end{pmatrix}.$$

Also find characteristic vectors corresponding to the characteristic values. Verify that $Ax = \lambda x$ in each case.

Since λ_1 and λ_2 are different (unless $R^2/L^2 = 4/LC$), the matrix Z that was described earlier will be nonsingular, and we can write

$$e^{At} = Z \begin{pmatrix} e^{\lambda_1 t} & 0 \\ 0 & e^{\lambda_2 t} \end{pmatrix} Z^{-1}$$

where Z and Z^{-1} are constant matrices (independent of t). This is the more transparent representation of e^{At} that we sought. We can now estimate the rate of decay of the *transient* part of the solution; for instance, when $t = 0$, then

$$e^{At} = Z\begin{pmatrix} 1 & 0 \\ 0 & 1 \end{pmatrix}Z^{-1} = ZIZ^{-1} = I$$

and in order to make each element (coefficient) in the matrix e^{At} smaller than, say, 0.01, we have only to satisfy the conditions

$$e^{u_1 t} < 0.01$$

$$e^{u_2 t} < 0.01$$

where u_1 and u_2 are the real parts of λ_1 and λ_2. These conditions will be satisfied for t such that

$$t > \frac{\ln 100}{-u_1}$$

$$t > \frac{\ln 100}{-u_2}.$$

In the case when the resistance R is smaller than $2\sqrt{L/C}$, then

$$u_1 = u_2 = -\frac{R}{2L}$$

and the steady state will have been "essentially reached" (within 1 percent) when

$$t > \frac{\ln 100}{\dfrac{R}{2L}} \approx 9.2\frac{L}{R}.$$

If we want to use $1/e$ instead of 0.01, then we need $t > 2L/R$. If $R > 2\sqrt{L/C}$, then λ_1 and λ_2 are both real and $e^{\lambda_1 t}$ decays more slowly than $e^{\lambda_2 t}$. In order to have $e^{\lambda_1 t} < 0.01$, we need

$$t > 9.2\left(\frac{1}{\dfrac{R}{L} - \sqrt{\dfrac{R^2}{L^2} - \dfrac{4}{LC}}}\right)$$

$$t > 9.2\frac{L}{R}\left(\frac{1}{1 - \sqrt{1 - \dfrac{4L}{R^2C}}}\right).$$

If R is close to $2\sqrt{L/C}$, then this is nearly $t > 9.2(L/R)$. If R is *much larger* than $2\sqrt{L/C}$, then

$$\left(1 - \frac{4L}{R^2C}\right)^{1/2} \approx 1 - \frac{1}{2}\frac{4L}{R^2C},$$

and we need

$$t > 4.6\, RC.$$

EXERCISES

1. Using the result of Exercise 1, page 186, find the steady-state solution

$$e^{At} \int_0^t e^{-At'} F(t') dt'$$

where

$$e^{At} = Z \begin{pmatrix} e^{\lambda_1 t} & 0 \\ 0 & e^{\lambda_2 t} \end{pmatrix} Z^{-1}$$

and

$$F(t) = \begin{pmatrix} 0 \\ \dfrac{2\pi f V_m \cos 2\pi f t}{L} \end{pmatrix}.$$

2. Suppose a certain biological system consists of one predator species and its prey. Let $N_1(t)$ represent the prey population at time t and $N_2(t)$ the predator. Consider the following mathematical model of the system:

$$\frac{dN_1}{dt} = a_1 N_1 (K - N_2)$$

$$\frac{dN_2}{dt} = a_2 N_2 (N_1 - R).$$

For positive constants a_1, a_2, K, R the prey will increase (in this model) if there are fewer than K predators, whereas the predators will increase only if there are more than R prey.

(a) Describe a *stable* (constant) steady-state situation.

(b) Suppose that $x_1 = N_1 - R$ and $x_2 = K - N_2$ are small (in absolute value).

What is the asymptotic (as t increases indefinitely) behavior of the *linearized* system obtained by neglecting terms higher than first degree in x_1 and x_2?

7.4   Solid-State Heat Conduction

Denote, by $u(t,x)$, the temperature at time t at a depth of x into a wall from the outside surface of the wall ($x = 0$). If the thickness of the wall is L and if the wall is made of a homogeneous material, then the *partial* differential equation of heat conduction

$$\frac{\partial u(t,x)}{\partial t} = k \frac{\partial^2 u(t,x)}{\partial x^2}$$

describes the behavior of temperature u as a function of time, t, and x, where k is the *coefficient of heat conduction* for the material in the wall (see Figure 7.2).

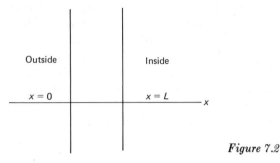

Outside

Inside

$x = 0$

$x = L$

x

Figure 7.2

Now consider the temperature near a point $\bar{x}$ within the wall. Suppose that at time t_0 the temperatures $u(t_0,x)$ for x near $\bar{x}$ give a curve such as shown in Figure 7.3. Near $\bar{x}$ the graph of $u(t_0,x)$ is shown as a *convex* curve, that is, with

$$\frac{\partial^2 u(t_0,\bar{x})}{\partial x^2} > 0.$$

The equation under consideration

$$\frac{\partial u}{\partial t} = k \frac{\partial^2 u}{\partial x^2}$$

implies that, at such a point $\bar{x}$, u will *increase* in time; that is, $\partial^2 u/\partial x^2 > 0$ implies $\partial u/\partial t > 0$, and so $u(t_0 + \Delta t,\bar{x}) > u(t_0,\bar{x})$ as long as $\partial u(t,\bar{x})/\partial t > 0$ for $t_0 \leq t \leq t_0 + \Delta t$. If, on the other hand, the curve is concave, with

$$\partial^2(u(t_0,\bar{x})/\partial x^2 < 0$$

at $\bar{x}$, then $\partial u(t_0,\bar{x})/\partial t < 0$ and $u(t,\bar{x})$ *decreases* as t increases (beyond t_0). If constant temperatures are maintained at both the inside and outside surfaces of the wall, say $u(t,0) = u_0$ and $u(t,L) = u_1$ for all t, then the temperature distribution within the wall will tend toward a steady-state *linear* distribution such that

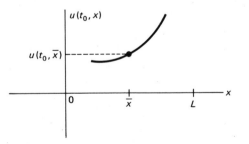

$u(t_0, x)$

$u(t_0, \bar{x})$

0

$\bar{x}$

L

x

Figure 7.3

$$u(t,x) = u_0 + \left(\frac{u_1 - u_0}{L}\right)x, \quad 0 \leq x \leq L$$

and both $\partial^2 u/\partial x^2(t,x)$ and $\partial u/\partial t(t,x)$ are zero for all x and t.

We can study the approach to the steady-state solution as follows. Suppose at time $t = 0$ we have $u_0 < u_1$ and that the temperature distribution within the wall is in the steady state as shown in Figure 7.4. Then, say at time $t_1 > 0$, suppose the outside surface temperature is suddenly increased to a new constant value $\bar{u}_0$ greater than u_1. At a time shortly after t_1, say t_2, the temperature distribution might look like the graph of $u(t_2,x)$ shown in Figure 7.5 as the solid curve.

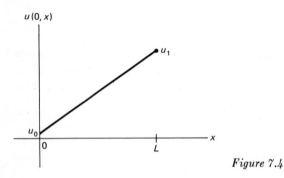

Figure 7.4

If we wait long enough we expect $u(t,x)$ to rise from the solid curve $u(t_2,x)$ shown in Figure 7.5 until the temperature distribution approaches the new steady state

$$u(t,x) = \bar{u}_0 + \left(\frac{u_1 - \bar{u}_0}{L}\right)x, \quad 0 \leq x \leq L$$

shown by the dashed line.

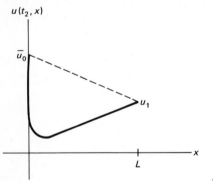

Figure 7.5

In the first of these two steady-state configurations, with $u_0 < u_1$, we would have to supply heat to the inside surface of the wall to maintain $u_1 > u_0$. In the second steady-state configuration, with $\bar{u}_0 > u_1$, we would have to remove heat from the inside surface of the wall to maintain a *lower* constant inside temperature.

We can attempt to estimate the *time* required to go from a temperature distribution such as shown in Figure 7.5 by the solid curve to the steady-state distribution shown by the dashed curve.

We could say that we have "essentially reached" the steady-state solution when, for each x in $[0,L]$, we have

$$\left[\bar{u}_0 + \left(\frac{u_1 - \bar{u}_0}{L}\right)x \right] - u(t,x) < 0.01 \left[\bar{u}_0 + \left(\frac{u_1 - \bar{u}_0}{L}\right)x - u(t_2,x) \right].$$

If we put

$$u = u(t,x) = Q_0(x) - e^{-\lambda t}Q(x)$$

with

$$Q_0(x) = \bar{u}_0 + \left(\frac{u_1 - \bar{u}_0}{L}\right)x,$$

then $\partial^2 Q_0 / \partial t = 0$ and we have

$$\frac{\partial u}{\partial t} = \lambda e^{-\lambda t}Q(x)$$

and

$$\frac{\partial^2 u}{\partial x^2} = -e^{-\lambda t}\frac{\partial^2 Q(x)}{\partial x^2}.$$

Substituting in the partial differential equation for heat conduction we obtain

$$\lambda e^{-\lambda t}Q(x) = -k e^{-\lambda t}\frac{\partial^2 Q(x)}{\partial x^2}$$

or

$$\frac{\partial^2 Q(x)}{\partial x^2} + \frac{\lambda}{k}Q(x) = 0.$$

Now $Q(x)$ is supposed to be a function of x alone, so the derivative $\partial^2 Q(x)/\partial x^2$ is the same as $d^2 Q(x)/dx^2$. We thus have a second-order, *ordinary* differential equation for $Q(x)$,

$$\frac{d^2 Q(x)}{dx^2} + \frac{\lambda}{k}Q(x) = 0$$

and, in order to satisfy $u(t,0) = \bar{u}_0$ and $u(t,L) = u_1$ we must require that

$$u(t,0) = -e^{-\lambda t}Q(0) + Q_0(0) = \bar{u}_0$$
$$u(t,L) = -e^{-\lambda t}Q(L) + Q_0(1) = u_1.$$
for all $t \geq t_2$

But $Q_0(0) = \bar{u}_0$, $Q_0(L) = u_1$ so that $Q(0) = Q(L) = 0$. Now the equation

$$\frac{d^2Q(x)}{dx^2} + \frac{\lambda}{k}\,Q(x) = 0$$

together with the boundary conditions

$$Q(0) = 0 \quad \text{and} \quad Q(L) = 0$$

has *solutions* of the form

$$Q(x) = Q_p(x) = \sin\left(\frac{2\pi p}{L}\right)x \qquad (p = 1,2,\ldots)$$

provided that $\lambda = \lambda_p$, where

$$-\left(\frac{2\pi p}{L}\right)^2 + \frac{\lambda_p}{k} = 0.$$

In other words, if we choose

$$\lambda = \lambda_p = k\left(\frac{2\pi p}{L}\right)^2,$$

then functions of the form

$$\varphi_p(t,x) = e^{-\lambda_p t}\sin\left(\frac{2\pi p}{L}\right)x \qquad (p = 1,2,\ldots)$$

satisfy the equation

$$\frac{\partial \varphi_p}{\partial t} = k\,\frac{\partial^2 \varphi_p}{\partial x^2}$$

and the boundary conditions $\varphi_p(t,0) = \varphi_p(t,L) = 0$ provided that

$$\lambda_p = k\left(\frac{2\pi p}{L}\right)^2.$$

If we write $u(t_2,x)$ (shown in Figure 7.5) as $u(t_2,x) = Q_0(x) - f(x)$, then $f(0) = 0, f(L) = 0$ and we can approximate $f(x)$ for $0 \leq x \leq L$ by a linear combination of the functions

$$[\varphi_p(t_2,x)] \qquad (p = 1,2,\ldots,N).$$

We can put

$$f(x) \approx \sum_{p=1}^{N} c_p\varphi_p(t_2,x)$$

where

$$\varphi_p(t_2,x) = e^{-\lambda_p t_2} \sin\left(\frac{2\pi p}{L}\right)x.$$

This amounts to a finite Fourier series with coefficients $c_p e^{-\lambda_p t_2}$ to be determined (see Section 5.7).

Without actually carrying out the determination of the coefficients $c_1, c_2, \ldots, c_N$, we can see that an approximate solution can be found of the form

$$u(t,x) = Q_0(x) - \sum_{p=1}^{N} e^{-\lambda_p t}c_p \sin\left(\frac{2\pi p}{L}\right)x$$

where

$$\lambda_p = k\left(\frac{2\pi p}{L}\right)^2.$$

We can verify this directly as follows. By taking partial derivatives of the expression for $u(t,x)$ just written, we have

$$\frac{\partial u(t,x)}{\partial t} = \sum_{p=1}^{N} \lambda_p e^{-\lambda_p t}c_p \sin\left(\frac{2\pi p}{L}\right)x$$

$$\frac{\partial^2 u(t,x)}{\partial x^2} = \sum_{p=1}^{N} \left(\frac{2\pi p}{L}\right)^2 e^{-\lambda_p t}c_p \sin\left(\frac{2\pi p}{L}\right)x$$

and so the expression satisfies

$$\frac{\partial u(t,x)}{\partial t} = k\,\frac{\partial^2 u(t,x)}{\partial x^2};$$

and, furthermore,

$$u(t,0) = Q_0(0) = \bar{u}_0$$

and

$$u(t,L) = Q_0(L) = u_1$$

as required; and, finally, when $t = t_2$, we have

$$u(t_2,x) = Q_0(x) - \sum_{p=1}^{N} e^{-\lambda_p t_2}c_p \sin\left(\frac{2\pi p}{L}\right)x$$

and the coefficients c_p, $(p = 1,2,\ldots, N)$ can be found to, at least approximately, satisfy this last *boundary condition* that expresses the initial shape of the curve $u(t,x)$ at $t = t_2$.

Without actually finding the coefficients $c_1, c_2, \ldots, c_N$, we can *estimate* the time required to "essentially reach" the steady-state solution $u(t,x) = Q_0(x)$ as the time it takes each term $e^{-\lambda_p t}$ to reach $e^{-\lambda_p t} \leq .01e^{-\lambda_p t_2}$, which is

$$t - t_2 \geq \frac{\ln 100}{\lambda p} \qquad p = 1, 2, \ldots, N.$$

The strongest condition occurs when $p = 1$ (as usual, the lowest frequency *waves* are the slowest to die out) and so our *estimate* of the time required to essentially reach the steady state is

$$t - t_2 = \frac{\ln 100}{k\left(\frac{2\pi}{L}\right)^2} \approx .12 \frac{L^2}{k}.$$

Recall that *essentially reach* was defined for convenience, in this example, as reducing to 1 percent a *given* difference between the steady-state solution and a given temperature distribution.

Appendix A

Closed form solutions

Sometimes, of course, it is possible to derive *closed-form* solutions to problems. As an illustration of the use of the *calculus* in deriving formulas (and for the information of the interested reader), we give here a derivation of a formula used earlier concerned with an earth-Mars flight.

As was stated in Section 1.4, the formula

$$V = \sqrt{\frac{2gM}{E(M + E)}} - V_0$$

gives an *approximation* to the minimum velocity with which a spaceship in free flight (rocket engines off) must leave the vicinity of the earth in order to reach Mars.

We now retrace the steps used to derive the formula, showing in this way the simplifying *assumptions* concerning the nature of the motion that were made along the way. First, *assume* that the earth and Mars move about the sun in circles with uniform (constant) angular velocity. This is only *approximately* the case. In order to derive the formula, assume it is *exactly* the case. Thus the formula has a first source of inaccuracy from this approximation. Furthermore, *assume* that the sun, earth, Mars, and spaceship all move within a single two-dimensional plane, as depicted in Figure A.1, with the sun, S, at the origin of the (x_1,x_2) coordinate system.

Assume that the earth and Mars orbit the sun in the direction indicated and that our spaceship will take a course as shown in the dotted elliptical curve. Denote the radius of the earth's orbit by E and that of Mars by M.

After a *spaceship* has left the earth's *immediate vicinity* (say 5 million miles away or so which would be outside the small circle about E in the

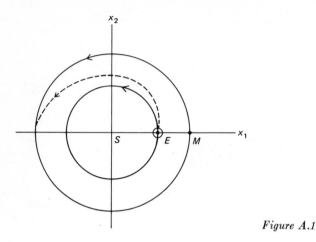

Figure A.1

figure), we can *assume* that it moves in powerless flight approximately according to Newton's *laws of motion* subject only to the gravitational attraction of the sun and its own momentum. If we denote the position in the figure of the spaceship by coordinates x_1, x_2, then Newton's laws of Motion state that, as functions of time, $x_1(t)$, $x_2(t)$ satisfy the differential equations

$$\ddot{x}_1 = -\frac{g}{r^2}\frac{x_1}{r}$$

$$\ddot{x}_2 = \frac{-g}{r^2}\frac{x_2}{r}.$$

We use the notation (for brevity)

$$\dot{x}_1 = \frac{dx_1(t)}{dt}, \quad \dot{x}_2 = \frac{dx_2(t)}{dt}$$

$$\ddot{x}_1 = \frac{d^2x_1(t)}{dt^2}, \quad \ddot{x}_2 = \frac{d^2x_2(t)}{dt^2}.$$

These differential equations incorporate the assumption that the acceleration of the spaceship $(\ddot{x}_1, \ddot{x}_2)$ due to the sun's gravitational attraction has a magnitude $\sqrt{\ddot{x}_1^2 + \ddot{x}_2^2}$ inversely proportional to the square of its distance, $r = \sqrt{x_1^2 + x_2^2}$, from the sun. The constant of proportionality g will be determined shortly. The attraction of the sun produces an *acceleration toward* the sun

$$\ddot{x}_1 = \left(-\frac{x_1}{r}\right)\sqrt{\ddot{x}_1^2 + \ddot{x}_2^2}$$

$$\ddot{x}_2 = \left(-\frac{x_2}{r}\right)\sqrt{\ddot{x}_1^2 + \ddot{x}_2^2}.$$

These equations together with

$$\sqrt{\ddot{x}_1^2 + \ddot{x}_2^2} = \frac{g}{r^2}$$

yield the differential equations for x_1 and x_2, which we wrote down, namely

$$\ddot{x}_1 = -\frac{g}{r^2}\frac{x_1}{r}$$

$$\ddot{x}_2 = -\frac{g}{r^2}\frac{x_2}{r}.$$

Let us start marking time at the instant when the spaceship leaves the "small" (5 million mile) circle about the earth. This circle *is* small relative to the orbits of the earth and Mars about the sun which we will take to have radii of approximately

$$E = 93 \cdot 10^6 \text{ miles}$$

and

$$M = 142 \cdot 10^6 \text{ miles}$$

respectively.

If we denote the position of the earth in the x_1, x_2 plane by X_1, X_2 and that of Mars by Y_1, Y_2, then the differential equations

$$\ddot{X}_1 = -\frac{g}{E^2}\frac{X_1}{E}$$

$$\ddot{X}_2 = -\frac{g}{E^2}\frac{X_2}{E}$$

$$\ddot{Y}_1 = -\frac{g}{M^2}\frac{Y_1}{M}$$

$$\ddot{Y}_2 = -\frac{g}{M^2}\frac{Y_2}{M}$$

are consistent with our assumption of circular orbits around the sun for earth and Mars.

Multiplying both sides of

$$\ddot{X}_1 = -\frac{g}{E^2}\frac{X_1}{E}$$

by $\dot{X}_1$ we get

$$\dot{X}_1\ddot{X}_1 = -\frac{g}{E^3}X_1\dot{X}_1.$$

Now

$$\frac{d}{dt}(\dot{X}_1^2) = 2\dot{X}_1\ddot{X}_1 \quad \text{and} \quad \frac{d}{dt}(X_1^2) = 2X_1\dot{X}_1$$

so we have

$$\frac{d}{dt}(\dot{X}_1^2) = \frac{-g}{E^3}\frac{d}{dt}(X_1^2).$$

In other words, $X_1(t)$ must satisfy

$$\frac{d}{dt}\left(\dot{X}_1^2 + \frac{g}{E^3}X_1^2\right) = 0.$$

If the derivative of a function is always zero, then the function must have a constant value. Therefore, for all t, we have

$$\dot{X}_1^2(t) + \frac{g}{E^3}X_1^2(t) = \dot{X}_1^2(0) + \frac{g}{E^3}X_1^2(0).$$

Since the earth moves with approximately constant angular velocity about the sun, we have

$$\dot{X}_1^2(t) + \dot{X}_2^2(t) = V_0^2$$

where

$$V_0 = \frac{2\pi E(\text{miles})}{1\ \text{year}} = \frac{2\pi \cdot 93 \cdot 10^6}{365 \cdot 24}\left(\frac{\text{miles}}{\text{hour}}\right)$$

or

$$V_0 = 66{,}500\ \text{mph.}\quad\text{(approximately)}.$$

Referring to Figure A.1, we have for $t = 0$, approximately

$$\dot{X}_1(0) = 0$$
$$X_1(0) = E$$

so

$$\dot{X}_1^2(t) + \dot{X}_1\frac{g}{E^3}X_1^2(t) = \frac{g}{E}\quad\text{(for any }t)$$

and, in particular, when t is such that $X_1(t) = 0$ and $X_2(t) = E$ (at the top of the earth's orbit), then $\dot{X}_2(t) = 0$ and $\dot{X}_1^2(t) = V_0^2$; so we obtain

$$V_0^2 = \frac{g}{E}.$$

Thus an approximate value of g is

$$g = EV_0^2 = 4.18 \cdot 10^{17}\left(\frac{\text{miles}^3}{\text{hour}^2}\right).$$

If we do the same calculation using the orbit of Mars, we obtain

$$g = M \cdot V_M^2 = 4.15 \cdot 10^{17}\left(\frac{\text{miles}^3}{\text{hour}^2}\right)$$

where

$$V_M = \frac{2\pi M}{1.9 \text{ years}} = \frac{2\pi \cdot 142 \cdot 10^6}{(1.9)(365)(24)} \frac{\text{miles}}{\text{hour}}$$

$$= 54{,}000 \text{ mph.} \quad \text{(approximately).}$$

This is in approximate agreement and we may take

$$g = 4.2 \cdot 10^{17} \left(\frac{\text{miles}^3}{\text{hour}^2}\right) \quad \text{(approximately).}$$

Getting back to the spaceship, we could send it off from the earth in the direction of the earth's orbit around the sun with a velocity $\dot{x}_2(0)$, which would then be the *sum* of $\dot{X}_2(0) = V_0$ and an excess over V_0 that we wish to minimize. We will make the further *assumption* that the way to do this is to give the spaceship just enough excess velocity (over V_0) so that it will move in an elliptical orbit which just touches and is tangent to the orbit of Mars at the point $(-M,0)$. We would have to time all this so that Mars itself gets to $(-M,0)$ just as the spaceship gets there.

If we go back to the equations of motion for the spaceship we have

$$\ddot{x}_1 = \frac{-g}{r^2} \frac{x_1}{r}$$

$$\ddot{x}_2 = \frac{-g}{r^2} \frac{x_2}{r}$$

where

$$r^2 = x_1^2 + x_2^2 \quad \text{and} \quad g = 4.2 \cdot 10^{17} \left(\frac{\text{miles}^3}{\text{hour}^2}\right).$$

If we multiply the first equation by $\dot{x}_1$ and the second by $\dot{x}_2$ and add the two equations, we obtain

$$\dot{x}_1 \ddot{x}_2 + \dot{x}_2 \ddot{x}_2 = \frac{-g}{r^2} \left(\frac{x_1 \dot{x}_1}{r} + \frac{x_2 \dot{x}_2}{r}\right).$$

The left-hand side can be written as

$$\dot{x}_1 \ddot{x}_1 + \dot{x}_1 \ddot{x}_2 = \frac{1}{2} \frac{d}{dt} (\dot{x}_1^2 + \dot{x}_2^2)$$

and the right-hand side can be written as

$$-\frac{g}{r^2} \left(\frac{x_1 \dot{x}_1}{r} + \frac{x_2 \dot{x}_2}{r}\right) = g \frac{d}{dt} \left(\frac{1}{r}\right).$$

To see this, notice that

$$\frac{d}{dt} r^2 = 2r \frac{dr}{dt} = \frac{d}{dt} (x_1^2 + x_2^2) = 2x_1 \dot{x}_1 + 2x_2 \dot{x}_2.$$

We can thus rewrite the summed equation above as

$$\frac{1}{2}\frac{d}{dt}\left(\dot{x}_1{}^2 + \dot{x}_2{}^2\right) = g\frac{d}{dt}\left(\frac{1}{r}\right),$$

which is the same as

$$\frac{d}{dt}\left[\frac{1}{2}\left(\dot{x}_1{}^2 + \dot{x}_2{}^2\right) - \frac{g}{r}\right] = 0.$$

Therefore, for all t, we have

$$\frac{1}{2}\left[\dot{x}_1{}^2(t) + \dot{x}_2{}^2(t)\right] - \frac{g}{r(t)} = \text{constant},$$

so we can write

$$\frac{1}{2}\left[\dot{x}_1{}^2(t) + \dot{x}_2{}^2(t)\right] - \frac{g}{r(t)} = \frac{1}{2}\left[\dot{x}_1{}^2(0) + \dot{x}_2{}^2(0)\right] - \frac{g}{r(0)}.$$

This equation, which expresses a *first integral* of the differential equations for $\ddot{x}_1$, $\ddot{x}_2$, is also an expression of the *law of conservation of energy*. Here $\frac{1}{2}\{\dot{x}_1{}^2(t) + \dot{x}_2{}^2(t)\}$ represents the *kinetic energy* of the spaceship at time t and $-g/r(t)$ represents its *potential energy*.

We can also derive an expression for the *conservation of angular momentum* for the sun-spaceship system by multiplying the equation for $\ddot{x}_1$ by x_2 and the equation for $\ddot{x}_2$ by x_1 and subtracting. We obtain

$$x_2\ddot{x}_1 - x_1\ddot{x}_2 = 0.$$

We then notice that

$$\frac{d}{dt}\left(x_2\dot{x}_1 - x_1\dot{x}_2\right) = x_2\ddot{x}_1 + \dot{x}_2\dot{x}_1 - \dot{x}_1\dot{x}_2 - x_1\ddot{x}_2$$

$$= x_2\ddot{x}_1 - x_1\ddot{x}_2 = 0.$$

Therefore, $x_2\dot{x}_1 - x_1\dot{x}_2 = \text{constant}$ for all t. So

$$x_2(t)\dot{x}_1(t) - x_1(t)\dot{x}_2(t) = x_2(0)\dot{x}_1(0) - x_1(0)\dot{x}_2(0).$$

This equation expresses the conservation of *angular momentum* for the orbit of the spaceship. It is another *first integral* of the differential equations.

We now have enough equations to put together the formula sought. Let T be the *time* at which the spaceship is supposed to reach Mars. Recall that we have (approximately)

$$x_1(0) = E, \quad x_2(0) = 0, \quad \dot{x}_1(0) = 0.$$

We wish to find $\dot{x}_2(0)$ such that $x_1(T) = -M$ and $x_2(T) = 0$. When this occurs, we will also have $\dot{x}_1(T) = 0$; so, using the conservation of angular momentum equation, we can write

$$M\dot{x}_2(T) = -E\dot{x}_2(0) \quad \text{or} \quad \dot{x}_2(T) = -\frac{E}{M}\,\dot{x}_2(0).$$

Using the conservation of energy equation, we can write (since $r(T) = M$ and $r(0) = E$),

$$\frac{1}{2}\,\dot{x}_2{}^2(T) - \frac{g}{M} = \frac{1}{2}\,\dot{x}_2{}^2(0) - \frac{g}{E}.$$

Substituting $-(E/M)\dot{x}_2(0)$ for $\dot{x}_2(T)$, we obtain

$$\frac{1}{2}\,\frac{E^2}{M^2}\,\dot{x}_2{}^2(0) - \frac{g}{M} = \frac{1}{2}\,\dot{x}_2{}^2(0) - \frac{g}{E},$$

and we can solve for $\dot{x}_2(0)$ to get

$$\dot{x}_2{}^2(0) = \frac{2g}{\left(\dfrac{E^2}{M^2} - 1\right)}\left(\frac{1}{M} - \frac{1}{E}\right)$$

$$= \frac{2gM}{E(E + M)}.$$

So the formula sought becomes

$$V = \dot{x}_2(0) - V_0 = \sqrt{\frac{2gM}{E(E + M)}} - V_0$$

which gives us a means for computing V from the values of g, M, E, and V_0.

Vectors, matrices, and linear transformations

A point x in the x_1, x_2 plane can be represented by a *two-dimensional vector*, that is, an ordered pair of numbers $x = (x_1,x_2)$ giving the two coordinates of the point. We say it is an *ordered* pair because, for instance, (1,2) does not represent the same point as (2,1) (see Figure B.1).

EXERCISES

1. Find the vector (point) (1,1) in Figure B.1.
2. If a given point in the plane is represented by the vector (x_1,x_2), then where is the point represented by $(-x_1,-x_2)$?
3. If (x_1,x_2) represents a point in the plane, then what is the *locus* of all points represented by vectors of the form (ax_1,ax_2) for any real number a?

An *n-dimensional vector* is a finite series (n-tuple) of numbers $(x_1,x_2,x_3,\ldots,x_n)$. The separate numbers x_1, x_2, . . . are called *components* of the vector. If x and y are n-dimensional vectors

$$x = (x_1,x_2, \ldots, x_n)$$

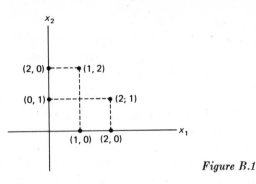

Figure B.1

and

$$y = (y_1, y_2, \ldots, y_n),$$

then a useful operation is that of taking the *inner product* defined by

$$(x, y) = x_1 y_1 + x_2 y_2 + \cdots + x_n y_n.$$

The *norm* of a vector (or at least *one* useful norm; there are many!) can be defined as (*Euclidean norm*)

$$\|x\| = (x, x)^{1/2}.$$

The *norm* of a vector is also called the *length* of the vector. Indeed the definitions just given correspond to the familiar length of a line segment in three-dimensional space. For, if (x_1, x_2, x_3) and (y_1, y_2, y_3) are the coordinates of two points x and y, then the line segment connecting x and y has length $\|x - y\|$, where $x - y = (x_1 - y_1, x_2 - y_2, x_3 - y_3)$; so $\|x - y\| = [(x_1 - y_1)^2 + (x_2 - y_2)^2 + (x_3 - y_3)^2]^{1/2}$.

Other operations involving vectors that are essential are *vector addition* and *scalar multiplication*. Addition of two vectors is defined by

$$(x_1, x_2, \ldots, x_n) + (y_1, y_2, \ldots, y_n) = (x_1 + y_1, \ldots, x_n + y_n).$$

Vectors are always added *componentwise*. Only vectors of the same dimensionality (same number of components) can be added. *Subtraction* of vectors is also defined *componentwise*.

Multiplication of a vector by a scalar (number) a is defined by

$$a(x_1, x_2, \ldots, x_n) = (ax_1, ax_2, \ldots, ax_n).$$

EXERCISES

1. Let $x = (1,0,1,2,2,-1)$ and $y = (0,1,1,-1,0,1)$ be two six-dimensional vectors. Carry out the following vector operations involving x and y:
   (a) Determine the *inner product* (x, y).
   (b) Find the norms $\|x\|$ and $\|y\|$.

(c) Find the vectors $x + y$ and $x - y$.

(d) Find $-2x$ and verify directly that $\|-2x\| = 2\|x\|$.

(e) Show that $\|ax\| = |a|\,\|x\|$ for any real number a and any vector x.

2. Write a computer program to find the inner product of two n-dimensional vectors.

A *matrix* is a rectangular array of numbers (called the *components* or *coefficients* or *elements* of the matrix). We can indicate which component of a matrix we are talking about by giving a pair of *indices* (integers) which give the number of the *row* (counting from the top down) and the *column* (counting from left to right) in which the component is to be found. The matrix

$$M = \begin{pmatrix} 3 & 2 & 2 & l \\ -1 & \pi & 0 & 4 \\ 0 & 0 & 0 & t \end{pmatrix}$$

is a 3×4 (three by four) matrix whose components, $M_{i,j}$, ($i = 1,2,$ or 3; $j = 1,2,3,$ or 4), are shown in the rectangular array given. For instance, $M_{1,1} = 3$, $M_{2,3} = 0$, $M_{3,4} = t$, and so on.

We can multiply a matrix, M, by a scalar, a, and obtain another matrix, aM, whose elements are $aM_{i,j}$. Two $k \times n$ matrices (k *may* be the same as n) may be added *componentwise*. A $k \times n$ matrix, A, may be multiplied by an $m \times k$ matrix, B, as follows. The product $C = BA$ has components given by

$$C_{i,j} = \sum_{p=1}^{k} B_{i,p} \cdot A_{p,j}.$$

The result C will be an $m \times n$ matrix. A $k \times 1$ matrix can be used to represent a k-dimensional vector (which is then called a *column vector*).

If B is an $m \times k$ matrix and a is a $k \times 1$ column vector, then Ba is an $m \times 1$ column vector with components

$$(Ba)_i = \sum_{p=1}^{k} B_{i,p} \cdot a_p.$$

EXERCISES

1. Carry out the following multiplications of a matrix times a vector (in *column* form).

(a) $\begin{pmatrix} 0 & 1 \\ -1 & 0 \end{pmatrix}\begin{pmatrix} 1 \\ 1 \end{pmatrix} = ?$

(b) $\begin{pmatrix} 1 & 2 & 2 \\ 0 & 1 & 0 \\ 1 & 0 & 1 \end{pmatrix}\begin{pmatrix} -1 \\ 3 \\ 1 \end{pmatrix} = ?$

(c) $\begin{pmatrix} 1 & 0 & 0 \\ 0 & 1 & 0 \\ 0 & 0 & 1 \end{pmatrix}\begin{pmatrix} 0 \\ 2 \\ 3 \end{pmatrix} = ?$

(d) $\begin{pmatrix} -1 & 0 & 1 \\ 3 & 2 & 2 \\ 1 & 3 & 1 \end{pmatrix}\begin{pmatrix} 0 \\ 1 \\ 0 \end{pmatrix} = ?$

2. Write a computer program to multiply an $n \times n$ matrix into an n-dimensional vector.

3. Multiply the matrices:

(a) $\begin{pmatrix} 0 & 1 \\ -1 & 0 \end{pmatrix}\begin{pmatrix} 0 & 1 \\ -1 & 0 \end{pmatrix} = ?$

(b) $\begin{pmatrix} 1 & 2 & 2 \\ 0 & 1 & 0 \\ 1 & 0 & 1 \end{pmatrix}\begin{pmatrix} -1 & 0 & 1 \\ 3 & 2 & 2 \\ 1 & 3 & 1 \end{pmatrix} = ?$

(c) $\begin{pmatrix} 1 & 0 & 0 \\ 0 & 1 & 0 \\ 0 & 0 & 1 \end{pmatrix}\begin{pmatrix} -1 & 0 & 1 \\ 3 & 2 & 2 \\ 1 & 3 & 1 \end{pmatrix} = ?$

4. Show that the matrix products

$$\begin{pmatrix} 1 & 1 \\ 0 & 1 \end{pmatrix}\begin{pmatrix} 0 & 0 \\ 1 & 1 \end{pmatrix} \text{ and } \begin{pmatrix} 0 & 0 \\ 1 & 1 \end{pmatrix}\begin{pmatrix} 1 & 1 \\ 0 & 1 \end{pmatrix}$$

are not the same.

5. Show that if A is any $n \times n$ matrix and I is the $n \times n$ *identity* matrix with components

$$I_{i,j} = \begin{cases} 1 & \text{if } i = j \\ 0 & \text{if } i \neq j, \end{cases}$$

then

$$IA = AI = A.$$

6. Write a computer program to find the product of two $n \times n$ matrices.

For each positive integer n, the set of all n-tuples or n-dimensional vectors is called Euclidean n-space, often written E^n (sometimes it is denoted by R^n, where $R^1 = R$ is the real line when $n = 1$). A function whose domain is E^n and whose range is in E^n is often called a *transformation* of E^n. Such a function can be thought of as moving (or *transforming*) each point to a new point. A *point*, x, in E^n is represented by an n-dimensional vector $x = (x_1, x_2, \ldots, x_n)$, giving the *coordinates* of the point.

A *linear* transformation, L, is a transformation with the property

$$L(ax + by) = aL(x) + bL(y)$$

for any vectors x and y and any scalars a and b.

Linear transformations of E^n can be represented by $n \times n$ matrices. Any point x in E^n has a representation $x = (x_1, x_2, \ldots, x_n)$, which can also be written as

$$x = x_1 e_1 + x_2 e_2 + \cdots + x_n e_n$$

where

$$e_1 = (1, 0, \ldots, 0)$$
$$e_2 = (0, 1, \ldots, 0)$$
$$\cdot$$
$$\cdot$$
$$\cdot$$
$$e_n = (0, 0, \ldots, 1)$$

are unit vectors on the coordinate axes. If L is a linear transformation of E^n, then

$$L(x) = x_1 L(e_1) + x_2 L(e_2) + \cdots + x_n L(e_n).$$

Now $L(e_1), L(e_2), \ldots, L(e_n)$ are again vectors and have representations in the form

$$L(e_1) = L_{1,1} e_1 + L_{2,1} e_2 + \cdots + L_{n,1} e_n$$
$$L(e_2) = L_{1,2} e_1 + L_{2,2} e_2 + \cdots + L_{n,2} e_n$$
$$\cdot$$
$$\cdot$$
$$\cdot$$
$$L(e_n) = L_{1,n} e_1 + L_{2,n} e_2 + \cdots + L_{n,n} e_n.$$

Thus $L(x)$ can be written as

$$L(x) = (L_{1,1} x_1 + L_{1,2} x_2 + \cdots + L_{1,n} x_n) e_1$$
$$+ (L_{2,1} x_1 + L_{2,2} x_2 + \cdots + L_{2,n} x_n) e_2$$
$$+ \cdot$$
$$\cdot$$
$$\cdot$$
$$+ (L_{n,1} x_1 + L_{n,2} x_2 + \cdots + L_{n,n} x_n) e_n.$$

Using the column-vector representation of points in E^n, we have

$$x = \begin{pmatrix} x_1 \\ x_2 \\ \cdot \\ \cdot \\ \cdot \\ x_n \end{pmatrix}, \quad L(x) = \begin{pmatrix} L_{1,1} x_1 + \cdots + L_{1,n} x_n \\ L_{2,1} x_1 + \cdots + L_{2,n} x_n \\ \cdot \\ \cdot \\ \cdot \\ L_{n,1} x_1 + \cdots + L_{n,n} x_n \end{pmatrix}.$$

Therefore, we can represent L by the $n \times n$ matrix

$$L = \begin{pmatrix} L_{1,1}L_{1,2}, \ldots, L_{1,n} \\ L_{2,1}L_{2,2}, \ldots, L_{2,n} \\ \cdot \\ \cdot \\ \cdot \\ L_{n,1}L_{n,2}, \ldots, L_{n,n} \end{pmatrix}$$

and we will have $L(x) = Lx$, where Lx denotes the column vector obtained by multiplying the column vector x by the matrix L. For example, suppose $L(x)$ is a linear transformation of the plane (E^2), with $L(e_1) = e_2$ and $L(e_2) = -e_1$, where $e_1 = (1,0)$ and $e_2 = (0,1)$, then the matrix representation of L is

$$L = \begin{pmatrix} L_{1,1}L_{1,2} \\ L_{2,1}L_{2,2} \end{pmatrix} = \begin{pmatrix} 0 & -1 \\ 1 & 0 \end{pmatrix}.$$

In column-vector notation, we have

$$e_1 = \begin{pmatrix} 1 \\ 0 \end{pmatrix} \quad \text{and} \quad e_2 = \begin{pmatrix} 0 \\ 1 \end{pmatrix}$$

so that

$$Le_1 = \begin{pmatrix} 0 & -1 \\ 1 & 0 \end{pmatrix}\begin{pmatrix} 1 \\ 0 \end{pmatrix} = \begin{pmatrix} 0 \\ 1 \end{pmatrix} = e_2$$

and

$$Le_2 = \begin{pmatrix} 0 & -1 \\ 1 & 0 \end{pmatrix}\begin{pmatrix} 0 \\ 1 \end{pmatrix}\begin{pmatrix} -1 \\ 0 \end{pmatrix} = -e_1.$$

EXERCISE

Suppose L is a linear transformation of E^2. If we know that

$$L[(1,1)] = (0,1)$$

and

$$L[(0,1)] = (2,1),$$

then what is the matrix representation of L?

Numerical solution of the N-body problem and the perihelion of mercury*

Local finite Taylor series expansion is shown to be an efficient numerical method for the study of the n-body problem. Application is made to the accurate determination of the motion of the perihelion of Mercury according to Newtonian celestial mechanics.

C.1 Introduction

Let x_i $(i = 1,2, \ldots, n)$ be the respective positions in three-dimensional Euclidean space of n bodies P_i with respective masses m_i. Let v_i be the veloc-

This Appendix has appeared as *Technical Report* #161, Computer Sciences Department, University of Wisconsin, October 1972, by R. E. Moore and D. Greenspan.

ity vector associated with P_i. Thus $d\mathbf{x}_i/dt = \mathbf{v}_i$. Newton's law of gravitation states that

$$\frac{d\mathbf{v}_i}{dt} = -\sum_{\substack{j=1 \\ j \neq 1}}^{n} \frac{Gm_j}{r_{ij}^3} (\mathbf{x}_i - \mathbf{x}_j) \qquad (i = 1,2,\ldots,n); \qquad \text{(C.1)}$$

where $r_{ij} = \|\mathbf{x}_i - \mathbf{x}_j\|$, G is the gravitational constant, and r_{ij} is assumed to be nonzero.

Recursion formulas will be developed for the determination of Taylor series coefficients to be used in the numerical solution of the system of equations (1.1). A procedure will be given for the automatic selection of variable step size to be used in successive finite (truncated) Taylor series expansions, and we will discuss the choice of order to be used. The resulting methods are applied to a detailed study of the motion of the perihelion of the planet Mercury over a 100-year period (the Julian century 1850.0–1950.0).

C.2   Recursion Formulas for Taylor Coefficients

For scalar and vector-valued functions of time we define the notation

$$(u)_k = \frac{1}{k!}\frac{d^k}{dt^k}u(t).$$

Notice that $(u)_0 = u(t)$. Thus $(u)_k$ is the kth-Taylor coefficient of u at t.

The following general recursion formulas are required (see R. E. Moore, 1966, Chapter 11).*

$$(u + v)_k = (u)_k + (v)_k$$
$$(u - v)_k = (u)_k - (v)_k$$
$$(uv)_k = \sum_{q=0}^{k} (u)_q(v)_{k-q}$$
$$(u^a)_k = \frac{1}{u}\sum_{q=0}^{k-1}\left(a - \frac{q(a+1)}{k}\right)(u)_{k-q}(u^a)_q$$
$$(u \neq 0, \; a \text{ is a real constant})$$
$$[(u)_1]_{k-1} = k(u)_k \qquad (k = 1,2,\ldots).$$

We will use the notation (u,v) for the inner product

$$(u,v) = u_1v_1 + u_2v_2 + u_3v_3.$$

*The formula (11–14) on p. 114 in References should read:

$$[P^a(x)]_j = \frac{1}{P(x)}\sum_{i=0}^{j-1}\left(a - \frac{i(a+1)}{j}\right)[P(x)]_{j-i}\,[P^a(x)]_i.$$

Using (C.2) we can derive the following recursion formulas for the Taylor coefficients of $\mathbf{x}_i$ satisfying (C.1):

$$(\mathbf{x}_i)_k = \frac{1}{k} (\mathbf{v}_i)_{k-1}$$

$$(r_{ij})_k = \frac{1}{r_{ij}} \left\{ \sum_{q=0}^{k-1} \left(1 - \frac{q}{k} \right) [(\mathbf{x}_i)_q - (\mathbf{x}_j)_q, \quad (\mathbf{x}_i)_{k-q} - (\mathbf{x}_j)_{k-q}] \right.$$

$$\left. - \left(\begin{matrix} 0, & \text{if } k = 1 \\ \sum_{q=1}^{k-1} \left(1 - \frac{q}{k} \right) (r_{ij})_q (r_{ij})_{k-q}, & \text{if } k > 1 \end{matrix} \right) \right\} \tag{C.3}$$

$$(r_{ij}^{-3})_k = \frac{1}{r_{ij}} \sum_{q=0}^{k-1} \left(-3 + \frac{2q}{k} \right) (r_{ij})_{k-q} (r_{ij}^{-3})_q$$

$$(\mathbf{v}_i)_k = -\frac{1}{k} \sum_{\substack{j=1 \\ j \neq i}}^{n} Gm_j \left(\sum_{q=0}^{k-1} [(\mathbf{x}_i)_q - (\mathbf{x}_j)_q] r_{ij}^{-3})_{k-1-q} \right) \qquad (k = 1, 2, \ldots).$$

Using (C.3) we fill out an array of values beginning with initial conditions for $\mathbf{x}_i$, $\mathbf{v}_i$ and calculate, for a given value of k, the quantities $(\mathbf{x}_i)_k$, $(r_{ij})_k$, $(r_{ij}^{-3})_k$, $(\mathbf{v}_i)_k$ from stored values for all previous values of k beginning with $k = 0$. For the quantities $(r_{ij})_k$ and $(r_{ij}^{-3})_k$ we need only use (C.3) when $i < j$. We can fill out the remainder of the array by using the relations

$$(r_{ji})_k = (r_{ij})_k$$
$$(r_{ji}^{-3})_k = (r_{ij}^{-3})_k.$$

The quantities $(r_{ii})_k$ and $(r_{ii}^{-3})_k$ do not appear. Using the coefficients obtained in this way, we have the following Kth-order (finite) Taylor series expansions

$$\mathbf{x}_i(t + h) = \sum_{q=0}^{K} (\mathbf{x}_i)_q h^q$$

$$\mathbf{v}_i(t + h) = \sum_{q=0}^{K} (\mathbf{v}_i)_q h^q \tag{C.4}$$

$$r_{ij}(t + h) = \sum_{q=0}^{K} (r_{ij})_q h^q.$$

Of course, we can find $r_{ij}^{-3}(t + h)$ from

$$r_{ij}^{-3}(t + h) = (r_{ij}(t + h))^{-3}.$$

The coefficients and series (C.4) are evaluated again at $t + h$ to carry the solution further. The method proceeds in this way in a step-by-step fashion.

C.3   Step Size and Order

Numerous theoretical and experimental studies (J. W. Daniel and R. E. Moore, 1970, Chapter 9; R. E. Moore, 1966, pp. 101–102) have shown that, in general, an approximately *optimal* choice for successive step sizes h_t in successive applications of a step-by-step Kth-order method, such as that given by (C.4), is as follows:

$$h_t = \left(\frac{N_0}{N_t}\right)^{1/K} h_0$$

where N_t is some functional representing an estimate of the coefficient of a leading error term of interest and h_0 is the initial step size. This has the effect of maintaining nearly constant *local truncation error*.

In *Interval Analysis* (R. E. Moore, Chapter 12) it is shown that an efficient choice for the order K in using Kth-order Taylor series to achieve approximately d decimal-digit accuracy locally is $K = d$.

Thus the most efficient order K for *maximum* possible accuracy in single precision on the UNIVAC 1108 (8 decimal digits) is about $K = 8$. In double precision on the UNIVAC 1108 (18 decimal digits) the most efficient choice of K for *maximum* possible accuracy is about $K = 18$. For $p \leq 18$ decimal-digit accuracy (locally, that is, at each step) we can take (in double-precision computation on the UNIVAC 1108)

$$h_t = \frac{1}{10(N_t(\mathbf{x}_i))^{1/p}} \tag{C.5}$$

in order to maintain a local truncation error of about $N_t(x_i)h_t{}^p = 10^{-p}$ in the term $\mathbf{x}_i(t)$. We can select, in this way, one of the n bodies in (C.1) for our particular attention. For $N_t(\mathbf{x}_i)$, we can use

$$N_t(\mathbf{x}_i) = \|(\mathbf{x}_i)_p\|. \tag{C.6}$$

After a number of trial calculations of the perihelion motion of Mercury using $K = 8$, $K = 12$, and $K = 18$, we settled upon the use of double precision (18 decimals) and $K = 12$ as a reasonable compromise between speed and accuracy for the main calculations to be reported on in the following section.

C.4   The Perihelion of Mercury

In 1947, G. M. Clemence (see Clemence, 1947, pp. 361–364) announced agreement of a prediction of Einstein's general theory of relativity concerning the motion of the perihelion of Mercury with long, existing discrepancies between observation and calculations based on Newtonian celestial mechanics.

We have undertaken carefully to verify or repudiate that claim by carrying out highly accurate computations based on a Newtonian model of motions of the solar system, using the exceptional power of modern computers. Indeed, the method we have used, based on the formulas presented in the previous sections, is particularly well suited for stored program computers, but hardly ideal for hand computation or desk computers, since it requires the intermediate storage of a rather large amount of data and the use of a fairly complicated set of formulas. The project to be described now was carried out using FORTRAN programming for the UNIVAC 1108 at the University of Wisconsin.

According to Chebotarev (1967, p. 28) accelerations relative to the center of the galaxy on objects in the solar system due to the motion of the solar system about the center of the galaxy amount to about $2 \cdot 10^{-9}$ in the units we will use, A. U. (*Astronomical Unit*) for a unit of distance, and 10 days ($= 1/36.525$ Julian years) for a unit of time. As will be seen from comparison with numbers to follow, we can neglect these accelerations and consider the solar system to be in an *inertial frame of reference* with its center of mass at the origin.

For our first computation, we started at 1951.0 and ran *backward* in time to 1850.0. As can be seen from (C.1) the solutions are invariant with respect to a simultaneous change in sign of t and v_i.

In what follows we will take X, Y, Z to be an inertial coordinate system with origin at the center of mass of the *solar system* which we will take to include only the sun, Mercury, Venus, Earth, and Jupiter since these are the major influences on the perihelion motion of Mercury. The X-axis will point toward the vernal equinox of 1951.0. The X, Y plane will be parallel to the plane of the earth's orbit around the sun at 1951.0. The Z-axis will be normal to the X, Y plane to make X, Y, Z a *right-handed* coordinate system. The equations of motion (C.1) are assumed to hold in the X, Y, Z coordinate system.

In a second, *heliocentric*, x, y, z coordinate system — with the sun at the origin and with the axes parallel to the corresponding axes in the X, Y, Z system — the earth is initially (at 1951.0) moving in the x, y plane. Again the X-axis is pointed toward the vernal equinox of 1951.0. In this coordinate system the oribital elements of the (relevant) planets are given by Krogdahl (1952, p. 95) as

Planet	a	e	i	Ω	ω	τ
Mercury	0.387	0.2056	7°0′14″	47°45′	28°57′	1950.993
Venus	0.723	0.0068	3°23′39″	76°14′	54°39′	1950.700
Earth	1.000	0.0167	0°	0°	120°6′	1951.008
Jupiter	5.203	0.0484	1°18′21″	99°57′	273°35′	1951.891

The *elements* given are: a, the semimajor axis; e, the eccentricity; i, the inclination of the planets orbital plane to the *ecliptic* (earth's orbital plane); Ω, the longitude of the ascending node; ω, the longitude (in the planet's orbital plane from the ascending node) of the perihelion point; and τ, the time of perihelion passage.

A set of transformations from the orbital elements to the rectangular heliocentric coordinates x, y, z is given (and was programmed) as follows (Roy, 1965, p. 106): the subscript i denotes the planet: $i = 1$, sun; $i = 2$, Mercury; $i = 3$, Venus; $i = 4$, earth; $i = 5$, Jupiter — to avoid confusion, we will denote *inclination* by I. For $i = 2, 3, 4, 5$, we have

$$n_i = (Gm_1 + Gm_i)^{1/2}a_i^{-3/2} \tag{C.7}$$

$$E_i - e_i \sin E_i = n_i(t - \tau_i) \tag{C.8}$$

$$l_{1,i} = \cos \Omega_i \cos \omega_i - \sin \Omega_i \sin \omega_i \cos I_i \tag{C.9}$$

$$m_{1,i} = \sin \Omega_i \cos \omega_i + \cos \Omega_i \sin \omega_i \cos I_i \tag{C.10}$$

$$n_{1,i} = \sin \omega_i \sin I_i \tag{C.11}$$

$$l_{2,i} = -\cos \Omega_i \sin \omega_i - \sin \Omega_i \cos \omega_i \cos I_i \tag{C.12}$$

$$m_{2,i} = -\sin \Omega_i \sin \omega_i + \cos \Omega_i \cos \omega_i \cos I_i \tag{C.13}$$

$$n_{2,i} = \cos \omega_i \sin I_i \tag{C.14}$$

$$b_i = a_i(1 - e_i^2)^{1/2} \tag{C.15}$$

$$x_i = a_i l_{1,i} \cos E_i + b_i l_{2,i} \sin E_i - a_i e_i l_{1,i} \tag{C.16}$$

$$y_i = a_i m_{1,i} \cos E_i + b_i m_{2,i} \sin E_i - a_i e_i m_{1,i} \tag{C.17}$$

$$z_i = a_i n_{1,i} \cos E_i + b_i n_{2,i} \sin E_i - a_i e_i n_{1,i} \tag{C.18}$$

$$r_i = (x_i^2 + y_i^2 + z_i^2)^{1/2} \tag{C.19}$$

$$\dot{x}_i = \frac{n_i a_i}{r_i} (b_i l_{2,i} \cos E_i - a_i l_{1,i} \sin E_i) \tag{C.20}$$

$$\dot{y}_i = \frac{n_i a_i}{r_i} (b_i m_{2,i} \cos E_i - a_i m_{1,i} \sin E_i) \tag{C.21}$$

$$\dot{z}_i = \frac{n_i a_i}{r_i} (b_i n_{2,i} \cos E_i - a_i n_{1,i} \sin E_i) \tag{C.22}$$

A few comments are in order concerning the transformation formulas. Equations (C.7) through (C.15) introduce auxiliary variables. Equation (C.8) defines E_i, the *eccentric anomaly* of planet i, implicitly. We use *Newton's method* to find E_i; putting $E_i^0 = n_i(t - \tau_i)$ where $t = 1951.0$ (after converting time to units of ten days) we iterated the formula

$$E_i^{(p+1)} = E_i^{(p)} - \frac{E_i^{(p)} - e_i \sin E_i^{(p)} - n_i(t - \tau_i)}{1 - e_i \cos E_i^{(p)}}$$

$p = 0, 1, 2, \ldots$, until* $|E_i^{(p+1)} - E_i^{(p)}| < 10^{-6}$, and then took $E_i^{(p+1)}$ for E_i.

*Giving $|E_i^{(p+1)} - E_i| \approx 10^{-12}$.

The resulting heliocentric initial conditions were (from computer printout).

i	x_i	y_i	z_i
2	$-.14271822.10^{-1}$	$.30803245$	$.26742133.10^{-1}$
3	$.43322016$	$-.58400374$	$-.33198056.10^{-1}$
4	$-.15591831$	$.97088167$	0
5	$.47831822.10^1$	$-.13874603.10^1$	$-.10192802$

i	$\dot{x}_i$	$\dot{y}_i$	$\dot{z}_i$
2	$-.33735568$	$-.20254059.10^{-2}$	$.30511267.10^{-1}$
3	$.16112397$	$.11982741$	$-.75903491.10^{-2}$
4	$-.17267800$	$-.27882252.10^{-1}$	0
5	$.20172708.10^{-1}$	$.76103702.10^{-1}$	$-.75267269.10^{-3}$

The gravitational constant was combined with the masses of the planets and the following values were used (see Clemence, 1947, p. 363.)

Sun: $\quad Gm_1 = .29591220.10^{-1}$ (A.U.^{3}/(10 days)^2)

Mercury: $\quad Gm_2 = .4931870.10^{-8}$

Venus: $\quad Gm_3 = .7252750.10^{-7}$

Earth: $\quad Gm_4 = .898364.10^{-7}$

Jupiter: $\quad Gm_5 = .2825234.10^{-4}$

To find the X, Y, Z coordinates with respect to the center of mass of the solar system, the following transformations are used:

$$GM = \sum_{i=1}^{5} Gm_i$$

$$\mathbf{x}_c = \frac{1}{GM} \sum_{i=2}^{5} Gm_i\mathbf{x}_i$$

$$\dot{\mathbf{x}}_c = \frac{1}{GM} \sum_{i=2}^{5} Gm_i\dot{\mathbf{x}}_i$$

$$X_i = \mathbf{x}_i - \mathbf{x}_c \quad (i = 1,2,3,4,5)$$

$$V_i = \dot{\mathbf{x}}_i - \dot{\mathbf{x}}_c$$

Let $\mathbf{X}_i = (X_i,Y_i,Z_i)$ and $\mathbf{V}_i = (VX_i,VY_i,VZ_i)$. The initial conditions in the inertial (X,Y,Z) coordinate system that result are as follows (at $t = 1951.0$):

Planet	i	X_i	Y_i	Z_i
Sun	1	$-.45629668.10^{-2}$	$.13218465.10^{-2}$	$.97299666.10^{-4}$
Mercury	2	$-.18834788.10^{-1}$	$.30935429$	$.26839433.10^{-1}$
Venus	3	$.42865720$	$-.58268189$	$-.33100756.10^{-1}$
Earth	4	$-.16048128$	$.97220352$	$.97299666.10^{-4}$
Jupiter	5	$.47786192.10^1$	$-.13861385.10^1$	$-.10183072$

Planet	i	VX_i	VY_i	VZ_i
Sun	1	$.19056125.10^{-4}$	$.72799117.10^{-4}$	$-.73143349.10^{-6}$
Mercury	2	$.33737474$	$.20982050.10^{-2}$	$-.30511998.10^{-1}$
Venus	3	$-.16110492$	$-.11975461$	$.75896176.10^{-2}$
Earth	4	$.17269706$	$.27955051.10^{-1}$	$-.73143349.10^{-6}$
Jupiter	5	$-.20153652.10^{-1}$	$-.76030902.10^{-1}$	$.75194126.10^{-3}$

The signs of the initial velocities were changed, replacing V_i by $-V_i$ ($i = 1,2,3,4,5$), and dt was replaced by $-dt$. The system (C.1) then was solved, using the method described in Sections C.2 and C.3, for the period from 1951.0 back to 1850.0. Numerical results are given in the following section.

In order to study the motion of the perihelion of Mercury during the century in question, the following items were computed and printed for each of 419 orbits of Mercury:

Time at perihelion T

Position of Mercury at perihelion (X_2,Y_2,Z_2)

Longitude of perihelion   THETA

(relative to the vernal equinox of 1951.0)

Perihelion distance $PD(=\|X_2 - X_1\|)$

As a check on accuracy, all ten known integrals to the n-body problem were computed and printed at each perihelion. These quantities, which remain constant for an exact solution, are as follows:

$$c_1 = \sum_{i=1}^{5} Gm_iX_i$$

$$c_2 = \sum_{i=1}^{5} Gm_iY_i$$

$$c_3 = \sum_{i=1}^{5} Gm_iZ_i$$

$$c_4 = \sum_{i=1}^{5} Gm_iVX_i$$

$$c_5 = \sum_{i=1}^{5} Gm_iVY_i$$

$$c_6 = \sum_{i=1}^{5} Gm_iVZ_i$$

$$c_7 = \sum_{i=1}^{5} Gm_i(X_i \cdot VY_i - Y_i \cdot VX_i)$$

$$c_8 = \sum_{i=1}^{5} Gm_i(Y_i \cdot VZ_i - Z_i \cdot VY_i)$$

$$c_9 = \sum_{i=1}^{5} Gm_i(Z_i \cdot VX_i - X_i \cdot VZ_i)$$

$$c_{10} = \sum_{i=1}^{5} \frac{1}{2} Gm_i(VX_i^2 + VY_i^2 + VZ_i^2) - \sum_{i<j} \frac{Gm_iGm_j}{R_{ij}}$$

where $R_{ij} = \|\mathbf{X}_i - \mathbf{X}_j\|$ as in (C.1).

The quantities c_1, c_2, c_3 are the coordinates of the center of mass and should remain zero. Similarly c_4, c_5, c_6 are the velocity components of the center of mass and should also remain zero. Now, c_7, c_8, c_9 are the components of total angular momentum of the system and should remain constant and equal to their initial values. Finally c_{10} is the total energy of the system and this should remain constant also and equal to its initial value.

The *time at perihelion* — when the planet Mercury is closest to the sun during an orbit — is determined as follows. We keep track of $(R_{12})_1$, which is the time derivative of the distance between Mercury and the sun. During the evaluation of the Taylor coefficients (C.3), $(R_{12})_q$ $(q = 1,2, \ldots, K)$, are computed, say at time t. Then we also evaluate

$$(R_{12})_1(t + h) = \sum_{q=1}^{K} (R_{12})_q \cdot q \cdot h^{q-1}.$$

We are going to use Newton's method to approximate h_1 such that $(R_{12})_1(t + h_1) = 0$. Therefore we also need $(d/dt)(R_{12})_1(t + h)$ or

$$2(R_{12})_2(t + h) = \sum_{q=2}^{K} (R_{12})_q(q)(q - 1)h^{q-2}.$$

We then find h_1 as follows. First, to distinguish perihelion from aphelion (farthest distance from the sun on Mercury's orbit) we test to see whether the following two conditions are both met (where h is the step size h_t determined by the method of Section C.3, using

$$N_t(\mathbf{X}_2) = |(X_2)_K| + |(Y_2)_K| + |(Z_2)_K|):$$

(1) $(R_{12})_1 < 0$
(2) $(R_{12})_1(t + h) > 0$.

When (1) and (2) occur together, there is an h_1 such that $(R_{12})_1(t + h_1) = 0$ and such that Mercury is at perihelion at time $t + h_1$. When (1) and (2) occur together, we put

$$h_1^{(0)} = h$$

and iteratively determine $h_1^{(p+1)}$ $(p = 0,1,2, \ldots)$ from

$$h_1^{(p+1)} = h_1^{(p)} - \frac{(R_{12})_1(t + h_1^{(p)})}{2(R_{12})_2(t + h_1^{(p)})}$$

until* $|h_1^{(p+1)} - h_1^{(p)}| < 10^{-8}$. Then we take $h_1 = h_1^{(p+1)}$ (the final iterate).

To find the *position of Mercury at perihelion*, we use the h_1 just determined and evaluate the finite series with vector coefficients

$$\mathbf{X}_2(t + h_1) = \sum_{q=0}^{K} (\mathbf{X}_2)_q h_1{}^q,$$

or, in component form,

$$X_2(t + h_1) = \sum_{q=0}^{K} (X_2)_q h_1{}^q$$

$$Y_2(t + h_1) = \sum_{q=0}^{K} (Y_2)_q h_1{}^q$$

$$Z_2(t + h_1) = \sum_{q=0}^{K} (Z_2)_q h_1{}^q.$$

To get the *perihelion distance*, we evaluate the three series for the components of

$$\mathbf{X}_1(t + h_1) = \sum_{q=0}^{K} (\mathbf{X}_1)_q h_1{}^q$$

and compute

$$\begin{aligned} PD &= \|\mathbf{X}_2(t + h_1) - \mathbf{X}_1(t + h_1)\| \\ &= \{[X_2(t + h_1) - X_1(t + h_1)]^2 + [Y_2(t + h_1) - Y_1(t + h_1)]^2 \\ &\quad + [Z_2(t + h_1) - Z_1(t + h_1)]^2\}^{1/2}. \end{aligned}$$

The *longitude* of *perihelion* (relative to the vernal equinox of 1951.0) is defined as "the sum of the two angles ω and Ω" mentioned near the beginning of this section (in connection with the table of orbital elements of the planets). Actually, at any given time the vernal equinox is defined as a directed half-line in the heliocentric coordinate system from the sun toward the *fixed stars* along which the plane of the earth's orbit (the *ecliptic*) and the plane of the earth's equator intersect. In reality, the vernal equinox is not fixed but moves relative to the heliocentric coordinate system in a slow arc due to the precession of the earth's axis of rotation.

We calculate ω, the angle in the plane of Mercury's orbit from the ascending note (where the plane of Mercury's orbit intersects the plane of the earth's orbit), and Ω, the angle between the x-axis and the ascending node, (see Chebotarev, 1967; Krogdahl, 1952; Roy, 1965).

Strictly speaking, this is not quite the *difference* between the correctly defined "heliocentric longitude of the perihelion of Mercury relative to the *moving* equinox" and the angle of precession of the equinox. But the differ-

---

*In the single-precision version we use $|h_1{}^{(p+1)} - h_1{}^{(p)}| < 10^{-6}$.

ence between $\omega + \Omega$ as we have defined it above and the correct version, $\omega + \Omega + (\Omega_t - \Pi_t - \Omega)$ due to motion of the ecliptic relative to the inertial frame, is $\Omega_t - \Pi_t - \Omega$, where Ω_t is the angle in the ecliptic plane from the moving equinox to the ascending node and Π_t is the angle from the moving equinox to the x-axis (equinox of 1951.0). This difference is of the order $1.6 \cdot 10^{-8}$ radian and does not affect our interpretation of the numerical results

We turn finally to a description of our method for computing the angles ω and Ω. At any fixed time t the plane of the earth's orbit around the sun is defined by the vectors

$$XE = \mathbf{X}_4 - \mathbf{X}_1$$

and

$$VXE = \mathbf{V}_4 - \mathbf{V}_1.$$

Similarly the plane of the orbit of Mercury is defined by

$$XM = \mathbf{X}_2 - \mathbf{X}_1$$

and

$$VXM = \mathbf{V}_2 - \mathbf{V}_1.$$

Thus a point V is in the intersection of these two planes (the line of nodes) if for some $\alpha, \beta, \gamma, \delta$

$$V = \alpha XE + \beta VXE = \gamma XM + \delta VXM.$$

Let us define a unit vector on the line of nodes. If we choose t, XM, VXM when Mercury is at perihelion, then $(XM, VXM) = 0$, and we will have $\|V\| = 1$ if

$$\gamma^2 \|XM\|^2 + \delta^2 \|VXM\|^2 = 1.$$

We will assume that XM, XE, and VXE are linearly independent; then

$$VXM = \bar{c}_1 XM + \bar{c}_2 XE + \bar{c}_3 VXE$$

for some $\bar{c}_1, \bar{c}_2, \bar{c}_3$ (since we are in a three-dimensional space). Then

$$(\gamma + \delta\bar{c}_1)XM + (\delta\bar{c}_2 - \alpha)XE + (\delta\bar{c}_3 - \beta)VXE = 0$$

and so $\gamma = -\delta\bar{c}_1$ (and also $\alpha = \delta\bar{c}_2$, $\beta = \delta\bar{c}_3$). We find, then, that

$$\delta = \left(\frac{1}{\|VXM\|^2 + \bar{c}_1^2\|XM\|^2}\right)^{1/2}.$$

and

$$V = \delta(VXM - \bar{c}_1 XM).$$

We need now to find $\bar{c}_1$. We can do this by taking the inner product of both sides of the equation

$$\bar{c}_1 XM + \bar{c}_2 XE + \bar{c}_3 VXE = VXM$$

with each of the three vectors XM, XE, and VXE. There results a system of three linear algebraic equations in $\bar{c}_1$, $\bar{c}_2$, $\bar{c}_3$ which we can solve for $\bar{c}_1$. A formal solution for $\bar{c}_1$, which we used in the computations, is

$$\bar{c}_1 = \frac{-\left(\dfrac{A_{12}}{A_{22}}\right)A_{24} - RR\left(A_{34} - \left(\dfrac{A_{23}}{A_{22}}\right)A_{24}\right)}{A_{11} - (A_{12}/A_{22})A_{12} - RR(A_{13} - (A_{23}/A_{22})A_{12})}$$

where

$$RR = \frac{A_{13} - \left(\dfrac{A_{12}}{A_{22}}\right)A_{23}}{A_{33} - (A_{23}/A_{22})A_{23}}$$

and where the A_{ij} are inner products that are defined as follows:

$$A_{11} = (XM,XM)$$
$$A_{12} = (XM,XE)$$
$$A_{13} = (XM,VXE)$$
$$A_{22} = (XE,XE)$$
$$A_{23} = (XE,VXE)$$
$$A_{24} = (XE,VXM)$$
$$A_{33} = (VXE,VXE)$$
$$A_{34} = (VXE,VXM).$$

Finally, we have

$$\omega = \text{arc cos}\left(\frac{(V,XM)}{\|V\| \cdot \|XM\|}\right)$$
$$= \text{arc cos}\left(\frac{-\bar{c}_1\|XM\|}{\|VXM - \bar{c}_1 XM\|}\right).$$

Actually we used the arc tangent function. If $\omega = \text{arc cos } Q$, then

$$\omega = \text{arc tan}\left(\frac{\sqrt{1 - Q^2}}{Q}\right)$$

so we take

$$\omega = \text{arc tan}\left(\frac{\sqrt{1 - Q^2}}{Q}\right)$$

where

$$Q = \frac{-\bar{c}_1\|XM\|}{\|VXM - \bar{c}_1 XM\|}.$$

The minus sign gives us the ascending node on the line of nodes.
For Ω we have

$$\Omega = \text{arc tan} \left(\frac{\sqrt{1 - U^2}}{U} \right)$$

where

$$U = \frac{VXM_1 - \bar{c}_1 XM_1}{\| VXM - \bar{c}_1 XM \|}.$$

and VXM_1, XM_1 are the *first components* of the vectors VXM and XM.
To study the motion of the perihelion of Mercury, we compute and printout the *change* in *longitude of perihelion*

$$\text{THETA} = \omega + \Omega - (\omega_0 + \Omega_0)$$

where ω_0 and Ω_0 are the initial values of ω and Ω for Mercury.

The results obtained for THETA versus a number of orbits of Mercury
(perihelion to perihelion) are shown in Figure C.1. A twelfth-order expansion was used at each step of the numerical integration. Thus we used
$K = 12$ in equations (C.4). Double precision was used (18 decimals). The
computing time was 53 minutes, 25 seconds. The ten integrals $c_1, c_2, \ldots, c_{10}$
all remained constant throughout to *at least* 8 decimals.

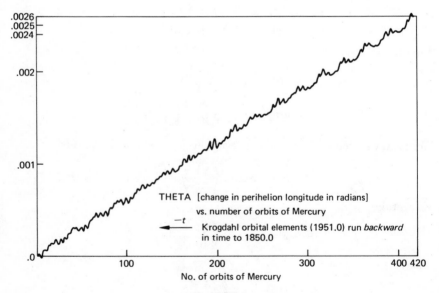

Figure C.1

A second set of initial conditions was then tried. This time we took the orbital elements given for the planets at 1850.0 by G. M. Clemence.*

Planet	i	a	e	i
Mercury	2	0.3870 986713	0.2056 0396	$7°0'7''$
Venus	3	0.7233 322169	0.0068 4458	$3°23'35.26''$
Earth	4	1.0000 00021	0.0167 7126	$0°$
Jupiter	5	5.2028 03945	0.0482 5382	$1°18'41.81''$

Planet	i	$\theta[=\Omega]$	$\Pi[=\omega+\Omega]$
Mercury	2	$46°33'12.24''$	$75°7'19.37''$
Venus	3	$75°19'47.41''$	$129°27'34.5''$
Earth	4	$0°$	$100°21'26.30''$
Jupiter	5	$98°55'58.16''$	$11°54'26.72''$

Planet	i	ϵ_0	$n\left(\dfrac{seconds}{year}\right)$	
Mercury	2	$323°11'23.53''$	538	1016.3893
Venus	3	$243°57'44.19''$	210	6641.4171
Earth	4	$99°48'18.56''$	129	5997.4496
Jupiter	5	$159°56'25.05''$	109	256.6395

The quantity ϵ_0 given here is defined by

$$\epsilon_0 - \omega - \Omega = n(t - \tau),$$

in terms of our previous variables. We had to convert the given values for n to our units of (radians/10 days). Calling the values in the table $\bar{n}$, we put

$$n = \frac{n}{(36.525)(57.295779)(3600)}.$$

Actually we did not need to solve for τ_i, but took instead $\epsilon_{0,i} - \omega_i - \Omega_i$ for the right-hand side of equation (C.8), which is used to determine E_i. We needed n_i as well for equations (C.20)–(C.22) to get $\dot{x}_i$, $\dot{y}_i$, $\dot{z}_i$.

This time we integrated forward in time from 1850.0 to 1950.0. Notice that we have, in this case, an inertial frame that puts the x, y plane in the *ecliptic of* 1850.0 and the x-axis is pointed toward the vernal equinox of 1850.0.

Again using double precision (18 decimals) and $K = 12$ (twelfth-order Taylor expansions at each integration step), we computed the perihelion

*G. M. Clemence, "First-Order Theory of Mars," *Astronomical Papers: American Ephemeris*, Vol. XI, Part II, United States Government Printing Office, Washington, 1949, pp. 231–232.

motion for 1850.0–1950.0. The results are shown in Figure C.2. There are only very minor differences between these and the previous results (using the 1951.0 elements and running backwards to 1850.0). The total computing time was 54 minutes, 27 seconds.

This time we are putting in orbital elements to several more decimal places and so the results are based on more accurate initial conditions. Again the integrals $c_1, c_2, \ldots, c_{10}$ remained constant to at least 8 decimals. Also shown on the figure is the value .002528, reported by Clemence (1947, p. 363), for the motion of the perihelion due to the Newtonian gravitational effects of Venus, earth, and Jupiter. He also reports a value for the *observed* motion relative to a fixed equinox (leaving out precession of the equinox) of .0027833. The effects of the other planets and solar oblateness (due to their Newtonian gravitational influence) adds up to about .0000485. The *general theory of relativity* predicts an increase of .000209 over what is predicted by Newtonian celestial mechanics. The sum of the Newtonian effects as computed by Clemence plus the relativistic term almost exactly equals the *observed* value.

All that *we* have computed, of course, is the motion of the perihelion due to the combined Newtonian influence of Venus, earth, and Jupiter. The

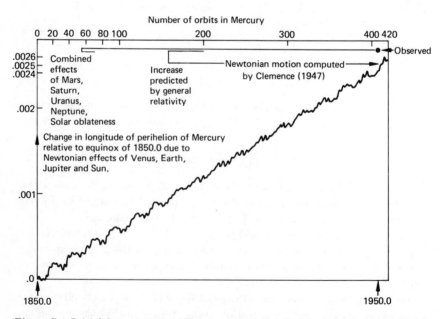

Figure C.2 Initial data at 1850.0 (Clemence, Astronomical Papers XI, part 2, p.232). Clemence (1947, p. 363) computes .002528 at 1950.0 for the Newtonian effects plotted in the graph; general relativity predicts an increase of .000209.

value .002528, given by Clemence for this contribution, is in excellent agreement with our results.

From our computed *times of perihelion*, we can derive values for the sidereal period of Mercury. The *average* period for the 415 orbits (perihelion to perihelion) during 1850.0–1950.0 according to our calculations is 87.9697213 days (taking one year = 365.25 days).

Finally, in Figure C.3 the perihelion distance (Mercury to sun at perihelion of Mercury) is shown (versus number of orbits of Mercury). There *seems* to be a *small rate of decrease* in addition to the fluctuations from orbit to orbit.

On the other hand, there are some extremely long-period phenomena involved here too. For instance, at the average rate of about .0025 (radians/100 years), it will take about 252,000 years for the longitude of Mercury's perihelion to make a complete revolution of 2π radians. Thus the perihelion distance may also have some very long-period fluctuations (see for example, Hagihara, 1961, pp. 95–158).

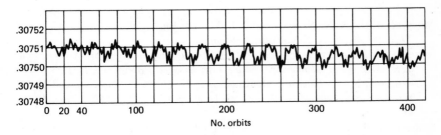

Figure C.3 Perihelion distance of Mercury vs. number of orbits of Mercury (1850.0–1950.0)

References

AHLBERG, J. H., E. N. NILSON, and J. L. WALSH. 1967. *The Theory of Splines and Their Applications*. New York: Academic.

AMERICAN MATHEMATICAL SOCIETY. 1968. "Some mathematical problems in biology," *Lectures on Mathematics in the Life Sciences*. Providence, R. I.

APOSTOLATOS, N., U. KULISCH, R. KRAWCZYK, B. LORTZ, K. NICKEL, and H. W. WIPPERMANN. 1968. "The algorithmic language triplex-Algol 60," *Num. Math.*, 11, 175–180.

BABUSKA, I., M. PRAGER, and E. VITASEK. 1966. *Numerical Processes in Differential Equations* New York: Wiley Interscience.

BOCHE, R. E. August 8, 1963. "An operational interval arithmetic," New York: Conf. paper, 63-1431, *IEEE*, pp. 1–13.

CHAI, A. S. 1967. "Statistical estimation of the effect of initial and roundoff errors in digital computation," Ph.D. thesis, Elect. Eng. Dept., Univ. of Wisconsin.

CHEBOTAREV, G. A. 1967. *Analytical and Numerical Methods of Celestial Mechanics*. New York: Elsevier.

CHENEY, W. 1960. "An example in differential equations: the n-body problem," *Amer. Math. Monthly*, 456–457.

CLEMENCE, G. M. October 1947. "The relativity effect in planetary motions," *Rev. of Mod. Physics*, 19, 4, 361–364.

COLLINS, G. 1960. "Interval arithmetic for automatic error analysis," *M & A*, 5, New York: IBM.

COLLATZ, L. 1960. *The Numerical Treatment of Differential Equations*. Berlin: Springer-Verlag.

DANTZIG, G. B. 1963. *Linear Programming and Extensions*. Princeton, N. J.: Princeton Univ. Press.

DANIEL, J. W. 1971. *The Approximate Minimization of Functionals*. Englewood Cliffs, N. J.: Prentice-Hall.

DANIEL, J. W., and R. E. MOORE. 1970. *Computation and Theory in Ordinary Differential Equations*. San Francisco: Freeman.

DEMPSTER, M. 1969. "Distributions in intervals and linear programming," in *Topics in Interval Analysis* (E. R. Hansen, Ed.), 107–127. New York: Oxford Univ. Press.

FADDEEV, D. K., and V. N. FADDEEVA. 1963. *Computational Methods of Linear Algebra*. San Francisco: Freeman.

FORSYTHE, G. E., and W. R. WASOW. 1960. *Finite Difference Methods for Partial Differential Equations*. New York: Wiley.

FORSYTHE, G. E., and C. B. MOLER. 1967. *Computer Solution of Linear Algebraic Systems*. Englewood Cliffs, N. J.: Prentice-Hall.

GIBB, A. 1961. "Algol 60 procedures for range arithmetic," *Tech. Rpt. No. 10*, Appl. Math. Stat. Lab., Stanford Univ. (Also *Comm. A, C, M.*, 4, July 1961.)

GINGERICH, O. 1964. "The computer versus Kepler," *Amer. Sci.*, 52, 218–226.

GIVENS, W. 1954. "Numerical computation of the characteristic values of a real symmetric matrix." *Rpt. ORNL 1574*, Oak Ridge, Tenn.

GREVILLE, T. N. E. 1969. "Spline functions and applications," *Orientation Lecture Series No. 8*. Math. Res. Ctr., Univ. of Wisconsin.

GOOD, D. I., and R. L. LONDON. June 1968. "Interval arithmetic for the Burroughs B-5500: four Algol procedures and proof of their correctness." *Comp. Sci. Dept. Tech. Rpt. No. 26*, Univ. of Wisconsin.

HAGIHARA, Y. 1961. "The stability of the solar system," in *The Solar System*, Vol. III (G. P. Kuiper and B. M. Middlehurst, Eds.), 95–158. Chicago: Univ. of Chicago Press.

HANSEN, E. R. 1965. "Interval arithmetic in matrix computations, Part I," *SIAM J. Num. Anal.* 2, 308–320.

HANSEN, E. R., and R. SMITH. 1967. "Interval arithmetic in matrix computations, Part II," *SIAM J. Num. Anal.* 4, 1–9.

HANSEN, E. R. (Ed.). 1969. *Topics in Interval Analysis*. New York: Oxford Univ. Press.

HENRICI, P. 1962. *Discrete Variable Methods in Ordinary Differential Equations*. New York: Wiley.

ISAACSON E., and H. B. KELLER. 1966. *Analysis of Numerical Methods*. New York: Wiley.

KIMELDORF, G., and G. WAHBA. 1971. "Some results on Tchebycheffian spline functions," *J. Math. Anal. Applic.* 33, 82–95.

KROGDAHL, W. S. 1952. *The Astronomical Universe*. New York: Crowell-Collier and Macmillan.

KUBA, D., and L. B. RALL. January 1972. "A UNIVAC 1108 program for obtaining rigorous estimates for approximate solutions of systems of equations." *Math. Res. Ctr. Tech. Summ. Rpt. 1168*, Univ. of Wisconsin.

LADNER, T. D., and J. M. YOHE. May 1970. "An interval arithmetic package for the UNIVAC 1108." *Math. Res. Ctr. Tech. Rpt. 1055*, Univ. of Wisconsin.

LOWANS, A. N., N. DAVIDS, and A. LEVENSON. 1943. "Table of zeros of the Legendre polynomials of order 1–16 and the weight coefficients for Gauss' mechanical quadrature formula," *Mathematical Tables and Other Aids to Computation*, 51–53.

Moore, R. E. December 1960. "Nonlinear two-point boundary problems with applications to the restricted three-body problem," *Lockheed Tech. Doc.: LMSD-
895025*, Palo Alto, Calif.

Moore, R. E. 1962. "Interval arithmetic and automatic error analysis in digital
computing," Appl. Math. & Stat. Lab., Stanford Univ. *Tech. Rpt. No. 25.* (Also
Ph.D. dissertation, Math. Dept., Stanford Univ., October 1962.)

Moore, R. E. 1966. *Interval Analysis.* Englewood Cliffs, N. J.: Prentice-Hall.

Moore, R. E. 1968. "Practical aspects of interval computation," *Aplikace Matema-
tiky*, **13**, 1, 52–92.

Nickel, K. 1968. "Anwendungen einer Fehlerschranken-Arithmetik," *Numerische
Mathematik, Differentialgleichungen, Approximationstheorie, ISNM*, **9**, 285–304.
Birkhauser Verlag: Basel und Stuttgart.

Nickel, K. 1969. "Triplex-Algol and its applications," in *Topics in Interval Analysis*
(E. R. Hansen, Ed.), 10–24. New York: Oxford Univ. Press.

Nickel, K. December 1971. "On the Newton method in interval analysis," *MRC
Tech. Rpt. No. 1136.* Univ. of Wisconsin.

Rall, L. B. (Ed.). 1965. *Error in Digital Computation*, Vols. I and II. New York:
Wiley.

Reiter, A. January 1968. "Interval arithmetic package (INTERVAL) for the
CDC 1604 and CDC 3600," *Math. Res. Ctr., Tech. Summ. Rpt. No. 794*, Univ.
of Wisconsin.

Richman, P. L. November 1969. "Variable-precision interval arithmetic," *Tech.
Mem., MM-69-1374-26*, Bell Telephone Labs.

Richman, P. L. December 1969. "Error control and the midpoint phenomenon,"
Tech. Mem., MM-69-1374-29, Bell Telephone Labs.

Richman, P. L. September 1972. "Automatic error analysis for determining precision," *Comm. A. C. M.*, **15**, 9, 813–817.

Robinson, S. M. August 1972. "Computable error bounds for nonlinear programming," *Math. Res. Ctr., Tech. Summ. Rpt. No. 1274*, Univ. of Wisconsin.

Roy, A. E. 1965. *The Foundations of Astrodynamics.* New York: Crowell-Collier and
Macmillan.

Schoenberg, I. J. 1967. "On spline functions," in Inequalities (O. Shisha, Ed.),
255–291. New York: Academic.

Steffensen, J. F. 1956. "On the restricted problem of three bodies," *Kgl. Danske
Videnskab., Mat. -fys. Medd.*, **30**, 18.

Stiefel, E. L., and G. Schiefele. 1971. *Linear and Regular Celestial Mechanics.*
New York: Springer-Verlag.

Talbot, T. D. June 1968. "Guaranteed error bounds for computed solutions of
nonlinear two-point boundary value problems," *Math. Res. Ctr. Tech. Summ.
Rpt. No. 875*, Univ. of Wisconsin.

Varga, R. S. 1962. *Matrix Iterative Analysis.* Englewood Cliffs, N. J.: Prentice-Hall.

Von Neumann, J., and H. Goldstine. 1947. "Numerical inversion of matrices of
high order," *Bull. A. M. S.*, **53**, 1021–1099.

Wilkinson, J. H. 1963. *Rounding Errors in Algebraic Processes.* Englewood Cliffs,
N. J.: Prentice-Hall.

WILKINSON, J. H. 1965. *The Algebraic Eigenvalue Problem*. New York: Oxford Univ. Press.

WILKINSON, J. H., and C. REINSCH. 1971. *Linear Algebra, Handbook for Automatic Computation*, Vol. II, New York: Springer-Verlag.

YOUNG, D. M. 1971. *Iterative Solution of Large Linear Systems*. New York: Academic.

Index

Index

Successive approximations, 28
Successive substitutions, 29
Summed version of quadrature formula,
    Gaussian quadrature, 132
    Simpson's rule, 128
    trapezoidal rule, 124

Tabular form, 85
Taylor series, 133–138, 160–163, 209–210
Terminal velocity of fall, 173–175
Time-dependent process, 173
Transient behavior, 173, 175, 177, 190–194
Trapezoidal rule, 124
Triangular inequality, 32
Two-dimensional vector, 202
Two-point boundary-value problem,
    163–168

Union, 75
Unit circle, 17
Unit vector, 105
Upper triangular form, 58

Variable of integration, 122
Variations, 46–51
Vector addition, 203
Vector differential equation, 150, 177
Vector space of functions, 102

Weierstrass' theorem, 111
Weights, 124, 126–127

Zero matrix, 180